Computational Optimization Techniques and Applications

Edited by Muhammad Sarfraz and Samsul Ariffin Abdul Karim

Published in London, United Kingdom

IntechOpen

Supporting open minds since 2005

Computational Optimization Techniques and Applications
http://dx.doi.org/10.5772/intechopen.87788
Edited by Muhammad Sarfraz and Samsul Ariffin Abdul Karim

Contributors
Jonah Lissner, Pushan Kumar Dutta, Umesh Prasad Verma, Subhra Mullick, Madhurendra Narain Sinha, Igor Sheremet, Mitsuru Uesaka, Katsuhiro Dobashi, Yuki Mitsuya, Jian Yang, Yoiichi Kusano, Aymen Lachheb, Lukman Abdurrahman, Gülenay Alevay Kılıç, Anand J. Patel, Pritam Pain, Goutam Kumar Bose, Michael Dinesh Simon, A.R.Kavitha Balaji, Abdolreza Asadi Ghanbari, Mousa Mohammadnia, Hossein Alaei, S.Abbas Sadatinejad, Paweł Kamiński, Maryam Mohd. Isa, Falih Salih Mahdi Salih Alkhafaji, Nasri Sulaiman, Wan Zuha Wan Hasan, Fatima Bouakkaz, Wided Ali, Makhlouf Derdour, Sabrina Guemadi, Lilia El Amraoui

First published in London, United Kingdom, 2021 by IntechOpen
IntechOpen is the global imprint of INTECHOPEN LIMITED, registered in England and Wales, registration number: 11086078, 5 Princes Gate Court, London, SW7 2QJ, United Kingdom
Printed in Croatia

British Library Cataloguing-in-Publication Data
A catalogue record for this book is available from the British Library

Additional hard and PDF copies can be obtained from orders@intechopen.com

Computational Optimization Techniques and Applications
Edited by Muhammad Sarfraz and Samsul Ariffin Abdul Karim
p. cm.
Print ISBN 978-1-83968-765-5
Online ISBN 978-1-83968-766-2
eBook (PDF) ISBN 978-1-83968-767-9

For EU product safety concerns:
IN TECH d.o.o., Prolaz Marije Krucifikse Kozulić 3, 51000 Rijeka, Croatia,
info@intechopen.com or visit our website at intechopen.com.

Meet the editors

Muhammad Sarfraz is a professor in the Department of Information Science, Kuwait University, Kuwait. His research interests include optimization, computer graphics, computer vision, image processing, machine learning, pattern recognition, soft computing, data science, and intelligent systems. Prof. Sarfraz has been a keynote/invited speaker at various platforms around the globe. He has advised/supervised more than 110 students for their MSc and Ph.D. theses. He has published more than 400 publications, including books, journal articles, and conference papers. His publications include around seventy books as author and editor. Prof. Sarfraz is a member of various professional societies. He is the chair and a member of the international advisory committees and organizing committees of various international conferences. He is also editor in chief and editor for various international journals.

Samsul Ariffin Abdul Karim is a senior lecturer at the Fundamental and Applied Sciences Department, Universiti Teknologi PETRONAS (UTP), Malaysia. He has been in the department for more than ten years. He obtained his BAppSc, MSc, and Ph.D. in Computational Mathematics and Computer Aided Geometric Design (CAGD) from Universiti Sains Malaysia (USM). He has twenty years of experience using Mathematica and MATLAB software for teaching and research activities. His research interests include curves and surfaces designing, geometric modeling, and wavelets applications in image compression and statistics. He has published more than 140 papers in journals and conferences as well as seven books, including two research monographs and three edited conference volumes, and thirty book chapters. He was the recipient of the Effective Education Delivery Award and Publication Award (Journal & Conference Paper), UTP Quality Day 2010, 2011, and 2012, respectively. He is a Certified WOLFRAM Technology Associate, Mathematica Student Level.

Contents

Preface

Computational optimization is an active and important area of study, practice, and research today. It covers a wide range of applications in engineering, science, and industry. It provides solutions to a variety of real-life problems in disciplines such as health, business, government, military, politics, security, education, and many more. Various problems can be transformed into optimization problems and then can be solved simply, accurately, and efficiently. This field is a source of revolutionizing, facilitating and enhancing the exchange of knowledge among researchers involved in both the theoretical and practical aspects. It emphasizes various topics including large-scale optimization, unconstrained optimization, constrained optimization, nondifferentiable optimization, combinatorial optimization, stochastic optimization, multi-objective optimization, linear programming, quadratic programming, parametric programming, complexity theory, automatic differentiation, approximations, error analysis, sensitivity analysis, theoretical analysis, evolutionary computing, surrogate-based methods, simulated likelihood estimation, support vector machines, and others. This comprehensive reference explores the developments, methods, approaches, and surveys of computational optimization in a wide variety of fields and endeavors. It focuses on optimization techniques, algorithms, analysis, applications, fields, nature of problems, and more.

This book compiles original and innovative findings on all aspects of computational optimization. It presents various examples of optimization including cost, energy, profits, outputs, performance, and efficiency. It also discusses different types of optimization problems like nonlinearity, multimodality, discontinuity, and uncertainty. The book also addresses various real-life applications of computational optimization in the fields of science, engineering, industry, health, business, government, military, information technology, and others. The book provides researchers, practitioners, academicians, military professionals, government officials, and other industry professionals with an in-depth discussion of the latest advances in the field. It consists of thirteen chapters in different areas of interest.

Kamiński starts the book with the first chapter on "Optimization Directions for Monitoring of Ground Freezing Process for Grzegorz Shaft Sinking." It discusses the sinking of the Grzegorz mineshaft in the Silesian Coal Basin in the United States, which is the first mineshaft sunk in the 21st century using a ground freezing method. Work carried out by the Shaft Sinking Company (PBSz S.A.) is characterized by a high level of innovativeness. Geophysical measurements were conducted to optimize the ground freezing process and its monitoring. Data gathered during research was used as a starting point for optimizing particular fields during the Grzegorz shaft sinking, as well as for use in future similar ventures. Proposed solutions might bring real improvements for the safety and effectiveness of the work as well as for economic factors. Conducted tests and analysis aimed at improving the monitoring of shape, size, and quality of the frozen rock mass column in a safe and reliable manner.

Chapter 2, "A Novel PID Robotic for Speed Controller Using Optimization Based Tune Technique" by Alkhafaji et al., investigates efforts to optimize coefficient gains, which is a significant issue for a proportional–integral–derivative (PID) controller. The authors propose massive tuning methods to resolve this problem, but there is little attention paid to optimize minimization time response significantly. The authors propose a technique to maximize optimization PID gains for the DC motor controller by combining a proper tuning method with a single-input and single-output (SISO) optimization toolbox using optimization-based tune (OBT) techniques that can be utilized for the greatest precision controller. A comparative study is carried out by applying five different tuning methods to obtain a proper tuning controller to then be combined with the SISO optimization toolbox. The utilized tuning methods include Robust Auto-Tune (RAT), Ziegler–Nichols (Z-N), Skogestad Internal Model Control (SIMC), Chien–Hrones–Reswick (CHR), and Approximate M-Constrained Integral Gain Optimization (AMIGO). The performance of all OBT tuning methods are analyzed and compared using the MATLAB/SISO tool environment, where efficiency is assessed based on time response characteristics (Ti) in terms of dead time (td), rise time (tr), settling time (ts), peak time (tp), and peak overshoot (Pos). The simulation results of the AMIGO-based proposal show a significant reduction time response measured in microseconds (μs). The novel feature of the proposed study is that it provides superior balancing between robustness and performance.

In Chapter 3, “Highway PC Bridge Inspection by 3.95 MeV X-Ray/Neutron Source,” Uesaka et al. develop portable 950 keV/3.95 MeV X-ray/neutron sources and apply them to the inspection of PC concrete thicker than 200 mm within a reasonable measuring time of seconds to minutes. T-girder-, Box- and slab- bridges are considered as part of the study. The authors suggest that it is time to begin X-ray transmission inspection for highway bridge (box) using 3.95 MeV X-ray sources in Japan. By obtaining X-ray transmission images of no-grout-filling in PC sheath and thinning of PC wires, they plan to carry out numerical structural analysis to evaluate the degradation of strength. Finally, the authors propose a technical guideline of nondestructive evaluation (NDE) of PC bridges taking into account both X-ray inspection and structural analysis. It has also been attempted to detect rainwater detection in PC sheath, and asphalt and floor slab by the 3.95 MeV neutron source. This is expected to be an early degradation inspection. Preliminary experiments are done on X-ray transmission imaging of PC wires and on-grout-filling in the same height PCs in 450–750-mm thick concretes. Moreover, the chapter also explains neutron backscattering detection of water in a PC sheath.

Chapter 4, “Incremental Linear Switched Reluctance Actuator,” by Lachheb and Amraoui describes that linear switched reluctance actuators are a focus of study for many applications because of their simple and robust electromagnetic structure. This is despite their lower thrust force density when compared with linear permanent magnet synchronous motors. This chapter deals with an incremental linear actuator, which has switched reluctance structure. It mentions different topologies of linear incremental actuators. The chapter places special focus on the switched reluctance linear actuator following the explanation of its operating principles. In addition, the authors develop an analytical model of the proposed actuator without taking account of the saturation in the magnetic circuit. Finally, the authors present control techniques that can be applied to the studied actuator.

In Chapter 5, “A Survey on Weapon Target Allocation Models and Applications,” Ghanbari et al. discuss Command and Control (C2), Threat Evaluation (TE), and Weapon Target Allocation (WTA). To build an automated system in this area after modelling TE and WTA processes, the models must be solved and an optimal solution must be found. This setting demands instantaneous operational planning and decision-making under inherent severe stress conditions. The associated responsibilities are usually divided among a number of operators and computerized decision support systems that aid these operators during the decision-making process. This chapter surveys the literature in the area of WTA systems with an emphasis on modelling and solving methods.

In Chapter 6, "A Review on Advanced Manufacturing Techniques and Their Applications" Patel and Kilic present a review on advanced manufacturing techniques and their applications. Advancement in manufacturing processes has drawn great interest from researchers and industry. It makes the process of manufacturing more productive and efficient. Advanced technology in manufacturing combines different manufacturing processes with similar objectives. This may include increasing material removal rate, improving surface integrity, reducing tool wear, reducing production time, and extending application areas. A combination of different processes is called a "hybrid" process. Hybrid processes open up new opportunities and applications for manufacturing various components that are not able to be produced economically by processes on their own. This chapter reviews the classification of current manufacturing processes based on the nature of the processing. It also includes reviews of existing and widely used manufacturing processes.

Chapter 7, "Stress-Strain Relationship: Postulated Concept to Understand Genetic Mechanism Associated with a Seismic Event" by Verma et.al, proposes a design methodology for monitoring earthquakes and detecting and tracking micro-seismic changes in the earthquake prediction system. The alert device includes sensors drastically different from current early warnings using the dozens of seismometers network across seismically active regions for the measurement of small acceleration signals directly. Specifically, first of all, it deals with the low-noise stage of the instruments measuring low-noise velocity signals. In this proposal, strain develops over time in the overlying stratum at a right angle to the applied shearing (max) stress. It obeys the internal friction of the stratum, available seismic energy, and laws of the stress-strain relationship. Using estimated energy (seismic), stress accumulation, the addition or subtraction in the strain rate due to stress developed, can be analyzed for a seismic event. This concept may lead to a better understanding of stress generation (build-up, transfer, and final drop). The chapter also proposes a methodology to identify the type of data to be used for spectral analysis in earthquake seismology and the type of instrument that can be used for data acquisition.

Chapter 8, "Nature Inspired Metaheuristic Approach for Best Tool Work Combination for EDM Process," by Bose and Pain discusses the use of electric discharge machining (EDM) with different types of tools like copper, aluminum, and brass while machining high-carbon high- chromium (HCHCr), hot die steel (HDS), and oil-hardened nitride steel (OHNS) workpiece material. The authors determine the most efficient tool material for different workpiece materials while satisfying the contradictory objectives of high material removal rate (MRR) and low tool wear rate (TWR). The experimental data are trained and validated using an artificial neural network (ANN). Finally, the results obtained through a genetic algorithm are hybridized with a fuzzy multi-criteria decision making (MCDM) technique to obtain a single parametric combination of the process control parameters that satisfies these two contradictory objectives simultaneously.

Global optimization in the 4D nonlinear landscape generates different types of particles, waves, and extremals of power sets and singletons. In Chapter 9, "Atomistic

Mathematical Theory for Metaheuristic Structures of Global Optimization Algorithms in Evolutionary Machine Learning for Power Systems," Lissner describes the atomistic mathematical theory for metaheuristic structures of global optimization algorithms in evolutionary machine learning for power systems. This chapter demonstrates the range of optimal problem-solving solution algorithms. Here, once, particles, or atoms of the ontological blueprint are generated inherently from the fractional optimization algorithms in metaheuristic structures of computational evolutionary development. These stigmergetics are applicable to incremental machine learning regimes for computational power generation and relay, and information management systems.

Chapter 10, "Ultrasonic Detection of Down Syndrome Using Multiscale Quantizer with Convolutional Neural Network" by Simon and Kavitha discusses echogenic intracardiac focus (EIF), which is one of the possible symptoms of Down syndrome (DS). In comparison to other symptoms like nasal bone hypoplasia and increased thickness in the nuchal fold, EIF is rare in DS. Hence, recommending pregnant women with EIF to undergo diagnostic processes like amniocentesis, chorionic villus sampling (CVS), and percutaneous umbilical cord blood sampling (PUBS) is not always the right choice, as these processes may result in serious adverse effects like miscarriage and uterine infections. This chapter presents a new ultrasonic method to detect EIF. It entails two stages: the training phase and the testing phase. The training phase aims at learning the features of EIF that can cause DS, whereas the testing phase classifies EIF into DS positive or DS negative based on the knowledge cluster formed during the training phase. A new algorithm, a multiscale quantizer with a convolutional neural network, is used in the training phase. An enhanced learning vector classifier is used in the testing phase to differentiate normal EIF from EIF causing DS. The performance of the proposed system is analyzed in terms of sensitivity, accuracy, and specificity.

Wireless sensor networks (WSNs) are used in diverse applications in the fields of military, agriculture, health care, medical monitoring, and more. The main issue of WSNs is energy consumption. Clustering with k-means is a successful technique for achieving a prolonged network lifetime using less energy. Chapter 11, "K-Means Efficient Energy Routing Protocol for Maximizing Vitality of WSNs" by Fatima et al. discusses the low-energy adaptive clustering hierarchy (LEACH) protocol, which is integrated with clustering where the choice of a number of clusters and their hierarchy uses the k-means method and the distance between nodes and residual energy. Clustering k-means gives the best partition with cluster separation. This chapter discusses related work using k-means to improve the vitality of WSNs. In addition, the chapter proposes an adaptation protocol. The simulation results using MATLAB show that the proposed protocol outperforms the LEACH protocol and optimizes the nodes' energy and the network's lifetime.

Today, Information Technology (IT) assists in providing main infrastructures for conducting business. Initially, IT emerged to automate business operations. As such, IT has been able to make business operations more effective due to its nature as a business enabler. Consequently, resource-based business efficiency is also the cornerstone of utilizing IT. Recently, IT has become over-the-top (OTT), where IT is the basic layer for conducting online business. In the current literature, the business operations rely on qualitative estimates and are not based on quantitative estimates. In Chapter 12, "Information Technology Value Engineering (ITVE)," Abdurrahman discusses Information Technology Value Engineering (ITVE). This chapter offers an IT value model whose value estimation can be done quantitatively using the Partial Adjustment Valuation (PAV) approach.

The book closes with Chapter 13, "Multi-Agent Implementation of Filtering Multiset Grammars" by Sheremet, who discusses the multi-agent implementation of filtering multiset grammars. This chapter is dedicated to the application of multi-agent technology to generate sets of terminal multisets (TMS) defined by filtering multiset grammars (FMG). The proposed approach is based on the creation of a multi-agent system (MAS), corresponding to specific FMG in such a way that every rule of FMG is represented by an independently acting agent. Such MAS provides a high-parallel generation of TMS and may be effectively used in any proper hardware environment. The chapter ends with a discussion of the directions for further developing the proposed approach.

Muhammad Sarfraz
Department of Information Science,
College of Life Sciences,
Kuwait University,
Kuwait

Samsul Ariffin Abdul Karim
Fundamental and Applied Sciences Department,
Universiti Teknologi PETRONAS (UTP),
Malaysia

Chapter 1

Optimization Directions for Monitoring of Ground Freezing Process for Grzegorz Shaft Sinking

Paweł Kamiński

Abstract

Grzegorz shaft is the first mine shaft sunk in 21st century in Silesian Coal Basin in USA of ground freezing method. Work carried out by Shaft Sinking Company (PBSz S.A.) is characterized by high level of innovativeness. Geophysical measurements were conducted to find directions of optimization of ground freezing process and its monitoring. Data gathered during research is a starting point for finding directions of optimization of particular fields during Grzegorz shaft sinking, as well as to be used in future similar ventures. Proposed solutions might have bring real improvements for safety and effectiveness of work and also for economic factors. Conducted tests and analysis aim at improvement of monitoring of shape, size and quality of frozen rock mass column in a safe and reliable manner.

Keywords: mine shaft, ground freezing, mine safety, geophysics measurements, monitoring methods in mining and civil engineering

1. Introduction. Grzegorz shaft: The first shaft in Silesian Coal Basin sunk using ground freezing method in 21st century

Grzegorz shaft is one of the biggest today's projects in Polish mining industry. Difficult conditions in its geological cross-section forced application of special shaft sinking method, which is ground freezing. This method was in common use in Silesian Coal Basin back in the days, but nowadays it is rarely used, mostly because of a small number of new shafts sunk. This venture was entrusted to Shaft Sinking Company (PBSz S.A.), part of the JSW Group and a leader of highly specialized mining services market in Poland. Company has got 75 years of experience in shaft sinking, including projects in extremely hard conditions, as well as in use of ground freezing method [1].

Application of typical shaft sinking method in case of Grzegorz shaft is impossible because of high rock mass' water accumulation and rock's low soaking resilience. Utilization of different methods, such as rock mass drainage and grouting, were analyzed, but expected low effectiveness effected in their abandonment. As the most effective, safe and reliable, ground freezing method was chosen for purpose of Grzegorz shaft sinking. Its essence is creating a column of frozen ground and it is realized by pumping freezing medium to boreholes. Mine shaft is sunk in such prepared rock mass using traditional methods, such as drill sand blasts. Column of frozen soils and rocks prevents shaft heading from flooding. It can also play a role of sidewall's support, as frozen rocks and soils are characterized by higher strength than those in natural state. Principles of ground freezing method and solution used for Grzegorz shaft sinking was described in details in following sections [1–3].

Grzegorz shaft was designed as a downcast, man and material shaft. Its inner cross-section is circle with a diameter of 7,5 m. Ordinate of surface level is +258,0 m and its depth is 870,0 m [3].

Shaft sinking is a huge and complicated venture, especially in terms of difficult geological conditions. However, Grzegorz shaft sinking is characterized by high level of innovativeness. One of the biggest innovations is application of the same head frame for both processes of sinking and regular operation of Grzegorz shaft. It is a first such case in Polish coal mining. Up to now, every mine shaft in Polish collieries was sunk using head frame of special construction, which were then disassembled and final head frame was built. But innovative way of thinking applies also to other areas of design and construction of Grzegorz shaft. Geophysical survey was conducted to determine new directions of optimization for processes of shaft sinking in a frozen rock mass and monitoring of ground freezing process [3].

2. Ground freezing method

The essence of special method of shaft sinking, which is ground freezing method is freezing of aquifers and then shaft sinking in frozen rock mass, using traditional methods, such as drills and blasts. It was primarily used in 1862 in England. Rock mass was frozen by freezing medium flow through a spiral pipe placed on a surface of quicksand's layer. In 1883 in Archibald mine, located near Schneidlingen, rock mass was frozen using technology similar to the one used nowadays. In Siberia's gold mines in 1940's shaft sinking method utilizing natural ground freezing was commonly used. It was then neglected, because of low effectiveness [4].

Low temperatures needed for freezing of soils and rocks around sunk shaft are obtained by heat of freezing medium transition from liquid to gas. The most common freezing medium is ammonia NH_3. Freezing boreholes are drilled around contour of the shaft. Distance between them is 0,9 to 1,2 m. They are equipped with casing pipe, so called freezing pipes, with diameter between 100 and 160 mm and pipes with diameter of 25–45 mm inside the freezing pipes. They are called inlet or inflow pipes, and are shorter than borehole depth (pipe do not reach borehole's bottom). Freezing medium is brought to a freezing ring on the surface, where it is distributed to freezing boreholes. It is then pumped into the boreholes through inflow pipes. Freezing medium flows between casing and inlet pipes, cooling rocks and soils via conduction. Constant rock mass cooling leads to freezing of water inside soils and rocks. Column of frozen ground develops around the freezing borehole. Such columns around numerous boreholes combine with each other, which effects in development of one cylinder of frozen ground around the outline of the shaft. This column of frozen soils and rocks prevents shaft heading from flooding and resists pressure of water and rock mass [4–11].

Various liquids can be pumped into freezing boreholes, such as aqueous solutions of calcium, sodium or magnesium chloride. All of them are characterized by low freezing temperature. The role of refrigerant is heat carrying, transferring it from rock mass to evaporator. It flows through freezing pipes in boreholes, collector and evaporator, where it is cooled down [4].

Strength of frozen rock or soil is higher than in natural state. The highest compressive strength characterizes frozen gravel and coarse-grained sand. Fine-grained sands and clays have lower compressive strength. Strength of frozen ground is also dependent on ice strength, which is related to ice grains' size and freezing time.

As rule of thumb, the faster freezing process the higher ice strength. Time of freezing depends on conductivity of casing pipes, freezing installation effectiveness and amount of heat to transfer [4–7].

Project of ground freezing has to be preceded by precise geological and hydrogeological measurements in the vicinity of designed shaft. In particular, it is important to identify hydrogeological conditions, such as number of aquifers, their depth and range, water pressure and its chemism [4, 8].

3. Geological conditions

3.1 Geological structure

Hydrogeological, geological and engineering conditions were determined on basis of data collected from boreholes G-8 and G-8bis, drilled specifically for this purpose. Stratigraphic profile in axis of the designed Grzegorz shaft consists of:

- **Quaternary formations** between 0,00 and 40,43 m – layers of sands, clay and aggregate;
- **Tertiary formations** between 40,43 and 114,43 m – layers of slit, clay, aggregate, claystone and limestone;
- **Triassic formations** between 114,3 and 234,93 m – layers of dolomite, limestone, clay, sandstone and mudstone;
- **Carboniferous formations** below 234,93 m – layers of sandstone, claystone and hard coal.

3.2 Hydrogeological conditions

Four aquifers with sixteen water bearing horizons are located in Grzegorz shaft profile. In quaternary formations there are two water bearing horizons, both with confined water table. It is fed by rainwater. Reservoir rocks are clays, sands and aggregates. Tertiary aquifer consists of single water bearing horizon with confined water table. It is also fed by rainwater. Reservoir rock is a limestone. Three water bearing horizons occur in Triassic aquifer, all of them with confined water table, also fed by rainwater. Reservoir rocks are dolomite, limestone, mudstone and sandstone. There are ten water bearing horizons in Carboniferous aquifer. Reservoir rock for all of them is sandstone. All horizons are also characterized by confined water table. There are fed by water infiltrating from upper stratigraphic layers.

Estimated water infiltration to the shaft heading from different aquifers varies between 0,057 to 0,926 m^3/min. Total expected water supply is equal 5,957 m^3/min.

3.3 Engineering conditions

According to observations made during drilling G-8bis borehole and laboratory tests of core sample there are seven different geotechnical zones, characterized with different geotechnical parameters.

It was found that there are extremely difficult geological conditions in zones I, III, V and VI, caused by low soaking resilience of rocks and high accumulation of water. Shaft sinking in such conditions is impossible without utilization of special methods, because there is a real threat of problems with sidewalls' stability.

Occurrence of these geotechnical zones enforced utilization of ground freezing method for Grzegorz shaft sinking.

4. Freezing installation and boreholes' construction

4.1 Freezing installation

Freezing installation consists of refrigeration plant (primary system) and secondary system and it is suited for cooling calcium chloride aqueous solution to −35°C [1, 3].

Primary system includes three freezing units, so called chillers PAC SAB 283 E eco, which freezing medium is ammonia. They are cooled by cooling towers Evapco AT 18-3 M14. Refrigeration plant consist also of isolated pipes for brine transfer, brine tank, discharge tank, water treatment station, eight pumps and armature. Model of refrigeration plant is presented on **Figure 1**.

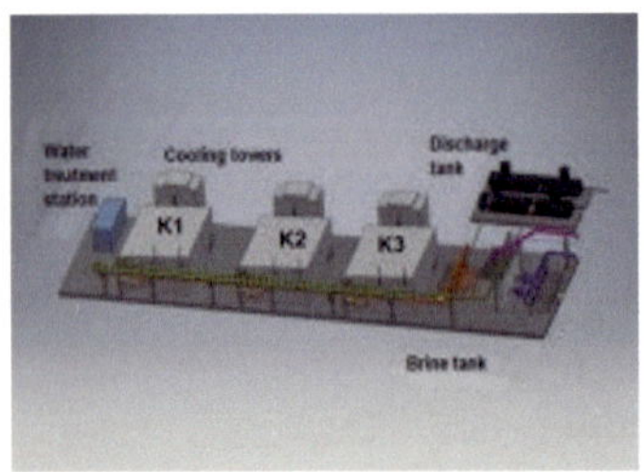

Figure 1.
Refrigeration plant [3].

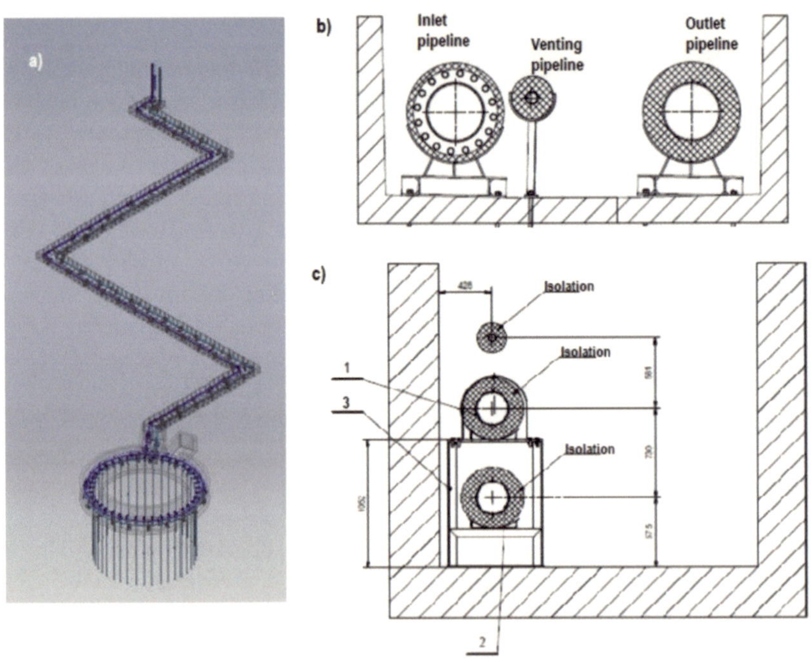

Figure 2.
Freezing channel [3]. a) Freezing channel model, b) cross-section of freezing channel between refrigeration plant and shaft; c) freezing channel around the shaft; 1: Inlet pipeline, 2: Outlet pipeline, 3: Rack.

Secondary system is basically a channel used to transfer coolant from refrigeration plant to freezing boreholes. The channel consists of two main parts, one of them is a circle channel around the shaft outline (containing freezing ring), the other one is a channel between freezing ring and refrigeration plant.

Freezing channel contains two pipelines with diameter of 350 mm – inlet and outlet pipelines and one venting pipeline with diameter of 65 mm. Around the shaft outline, pipelines are circle-shaped and form so-called freezing ring. Freezing boreholes' pipes are connected directly to the freezing ring. Model of secondary system of freezing installation and cross-sections of channels are shown on **Figure 2**.

4.2 Freezing boreholes

For purpose of Grzegorz shaft sinking, 40 freezing and 3 control boreholes were designed and drilled. They are marked with numbers from M1 to M40 (freezing boreholes) and from T1 to T3 (control boreholes; one of them is previously drilled borehole G-8bis). Control boreholes are located on common axis, out of the freezing boreholes' circle. Their task is to inspect shape and size of frozen ground column. Depth of all boreholes is equal 485 m (bottom 10 meters are filled with concrete) [12].

Distance between freezing boreholes varies in practice between 0,9 and 1,3 m. It is determined according to:

- required size of frozen ground cylinder,
- economic factors,
- time required.

In case of Grzegorz shaft sinking distance between freezing boreholes was calculated at about 1,25 m. Parameters of freezing boreholes are presented in **Table 1**.

Construction of freezing boreholes:

- 0–2,0 m – lining pipe Ø508 mm (20″)
- 2,0–40,0 m – borehole's diameter Ø311 mm – 12 1/4″ (rotary drill bit Ø311 mm)
- 40–485 m – borehole's diameter Ø216 mm – 8 1/2″ (rotary drill bit Ø216 mm)

Figure 3 presents arrangement of freezing and control boreholes and Grzegorz shaft's cross-section. Description of shaft's elements represents state of construction at the moment of beginning of drilling.

No.	Parameter	Value
1.	Number of freezing boreholes	40 pcs
2.	Distance between boerholes	1,25 m
3.	Diameter of freezing boreholes circle	16 m
4.	Depth of freezing boreholes	485 m (including 10 m of concrete)
5.	Casing pipes diameter	Ø139,7 mm × 8,0 mm
6.	Inflow pipes diameter	Ø75,0 × 4,5 mm
7.	Refrigerant	$CaCl_2$ aqueous solution

Table 1.
Parameters of freezing boreholes.

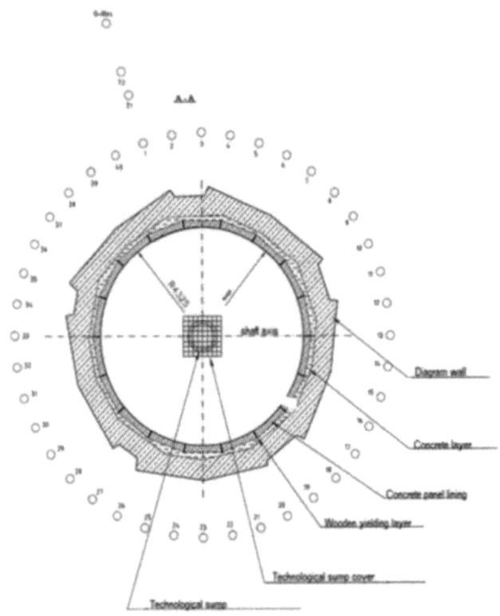

Figure 3.
Arrangement of freezing and control boreholes and cross-section of Grzegorz shaft [1].

5. Geophysical measurements

Monitoring of phenomena occurring in frozen rock mass is crucial for safety of the whole process of shaft sinking using ground freezing method. It is not an easy task, though. Number of geophysical measurements were conducted on Grzegorz shaft's construction plant in order to optimize process of ground freezing and monitoring of frozen rock mass. On a basis of gathered data, practical aspects of presented methods were analyzed.

Potential measuring methods assumed possible to use for Grzegorz shaft monitoring were [13]:

- vertical seismic profiling,
- transmission tomography,
- seismic interferometry,
- surface wave analysis,
- reflection seismology.

Surface waves analysis allows estimation of velocity of transverse wave behind the shaft lining, within few meters. Utilization of reflection seismology helps to anticipate geological faults and some of the parameters of rock mass. Especially prior recognition of faults plays critical role in safety of miners working in the shaft's heading, as well as for economic factors. Application of proposed methods has to be carefully considered and tested in real-life condition. Unfortunately, it was impossible, because of the small depth (20 meters) of Grzegorz shaft during measurement session. Initial assumptions involve utilization of geophones in short distance from shaft's heading [13].

An objective of tomography test is estimation of 2D wave velocity field between adjacent boreholes. Tests were carried out using vibration source of special construction, made especially for purpose of these tests. This device is presented on **Figure 4**. Two tests were made, first of them with utilization of probe consisting of

Figure 4.
A view of vibration source prototype.

four hydrophones with own frequency of 500 Hz and another one with using two probes equipped with three hydrophones with own frequency of 100 Hz [13].

The results obtained prove that monitoring of the frozen rock mass column using probes and vibration source is possible. High financial expenditures needed for vibration source construction (prototype used for tests is suited for use in control boreholes, final vibration source for use in freezing boreholes has to be constructed) and problematic technology of testing caused neglecting of further development of such solution [13].

Utilization of scattered waves allows monitoring of deformational phenomena occurring in rock mass. The appearance of cracks, rifts and microfractures affects velocity of wave propagation and its amplitude. What is even more important in case of ground freezing is high sensitivity of coda wave velocity for (even small) temperature changes in rock mass [13–16].

Utilization of seismic interferometry was proposed for purpose of monitoring of phenomena occurring in frozen rock mass caused by temperature changes in the vicinity of the Grzegorz shaft. Sources of seismic noise, which is needed for such measurements, are operating freezing boreholes. This method of monitoring provides near real-time information.

Tests were conducted to prove correctness of initial assumptions. Measurements were carried out using two probes consisting of three hydrophones with own frequency of 100 Hz each. **Figure 5** presents 5-minute record of seismic noise. Channels 0, 1 and 2 represent hydrophones located in control borehole T2, channels 3, 4 and 5 represent hydrophones used in borehole T1, located closer to freezing boreholes. High amplitude values about 13:02 are caused by tests of vibration source used in the tomography measurements.

From the practical point of view, the most relevant is the period of stationary seismic noise. It is probably an effect of freezing boreholes operation and its frequency varies between 10 and 200 Hz **Figure 6**. Presents record and spectrogram of channel 0. Spectrogram's scale upper limit is 300 Hz.

On the basis of scattered wave interferometry correlation functions, representing estimated scattered wave propagation between hydrophones, were obtained. Further tests and analysis are required for purpose of utilization of this method for real life application, including tests covering long time period of measurement.

To sum up, coda wave interferometry method can be used in practice, for monitoring of ground freezing process and become convenient and reliable monitoring method. High sensitivity for temperature changes can prevent potential unexpected failures, which are real threat for miners working in the shaft heading and for shaft

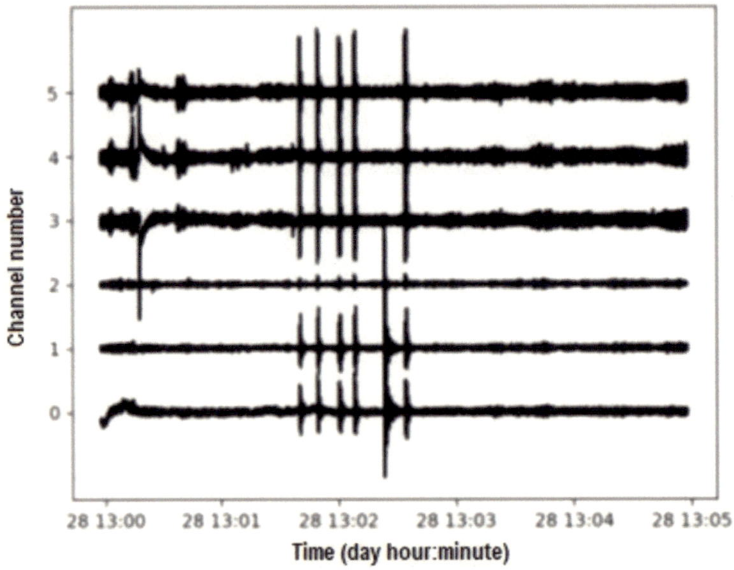

Figure 5.
Example of seismic noise record.

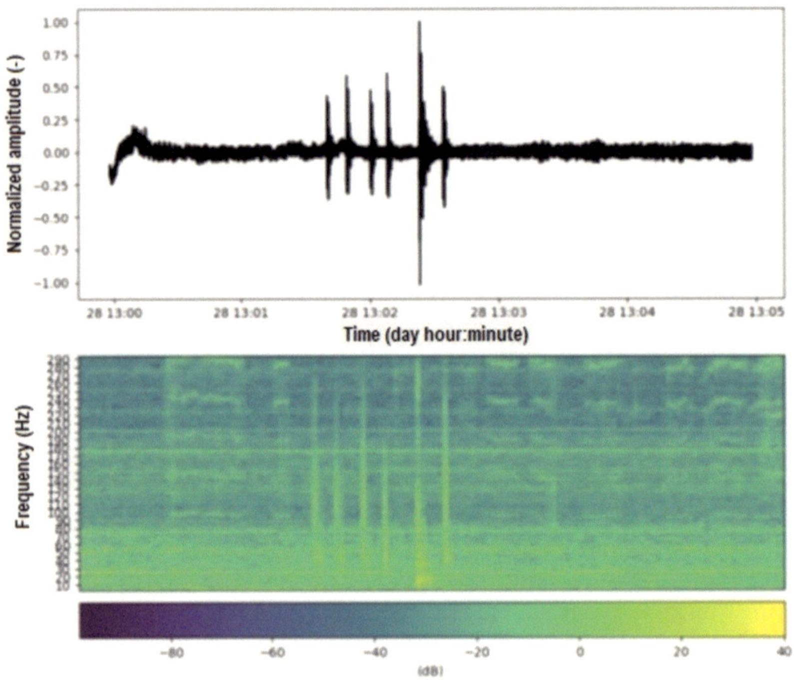

Figure 6.
Record and spectrogram of channel 0.

itself. However, this method has to be tested in long period tests, spanning for several days, as well as in situation of controlled stoppage of operation of freezing borehole, to check if the monitoring system work well in hazardous situation [13].

The most promising data was obtained in a test of vertical seismic profiling. Tests were conducted using device with parameters presented in **Table 2**. The source of seismic wave was generated on the surface in the near vicinity of a borehole by a 5 kg sledgehammer [13, 17, 18].

Seismograph	Geode 24CH Geometrics
Probe	4 hydrophones with own frequency of 500 Hz
Distance between hydrophones	0,2 m
Measuring step	0,6 m
Amount of hits	1 for each probe position (each 0,6 m)
Record time	0,5 s
Frequency	4000 Hz
Depth	0–200 m

Table 2.
Parameters of seismic profiling record [13].

Figure 7 presents results of seismic record initial analysis for single measurement in the borehole T1. Traces were filtered in frequency range between 150 and 800 Hz and normalized to common average. Regular amplitudes, effecting from wave propagation, can be observed in the figure.

Basing on data obtained from profiling boreholes T1, T2 and T3, a collective map of wave velocity distribution was computed. It is limited to a depth where data obtained is characterized by high quality. Resulting map is shown in **Figure 8**.

There is a relationship between temperature (data collected at the beginning of ground freezing process), presented in **Figure 9**, and velocity distribution, shown in **Figure 10**. Data obtained from borehole T1 was chosen for presentation, because of the shortest distance to freezing boreholes.

Temperature and wave velocity were also compared on one graph, presented in **Figure 11**. Relationship between them was estimated.

Non-linear, polynomial function with constant parameters was obtained. Function was suited with high correlation factor, equal 0,82. It is clear that there is a relationship between temperature and velocity.

Consequently, graphical correlation of frozen rock mass column development, temperature curve and map of velocity distribution was prepared. It is shown in **Figure 12**. Blue sections represent temperature and velocity drops, red color indicates temperature and velocity rises. Relationship is clear [13].

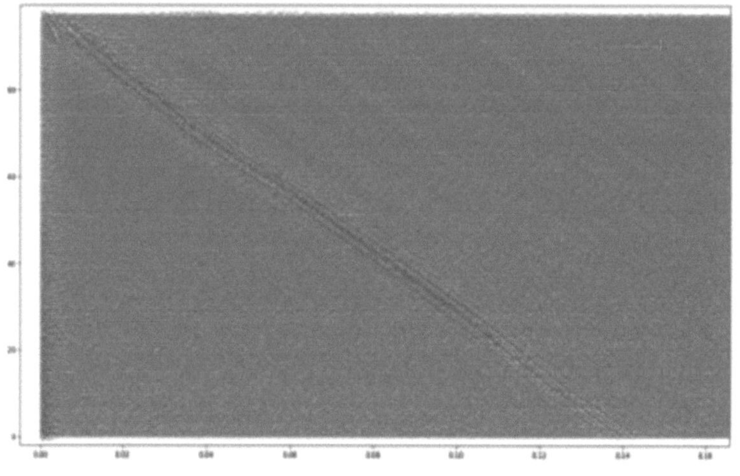

Figure 7.
Example of seismic record in the borehole T1; vertical axis – Numbers of traces, horizontal axis – Time in ms.

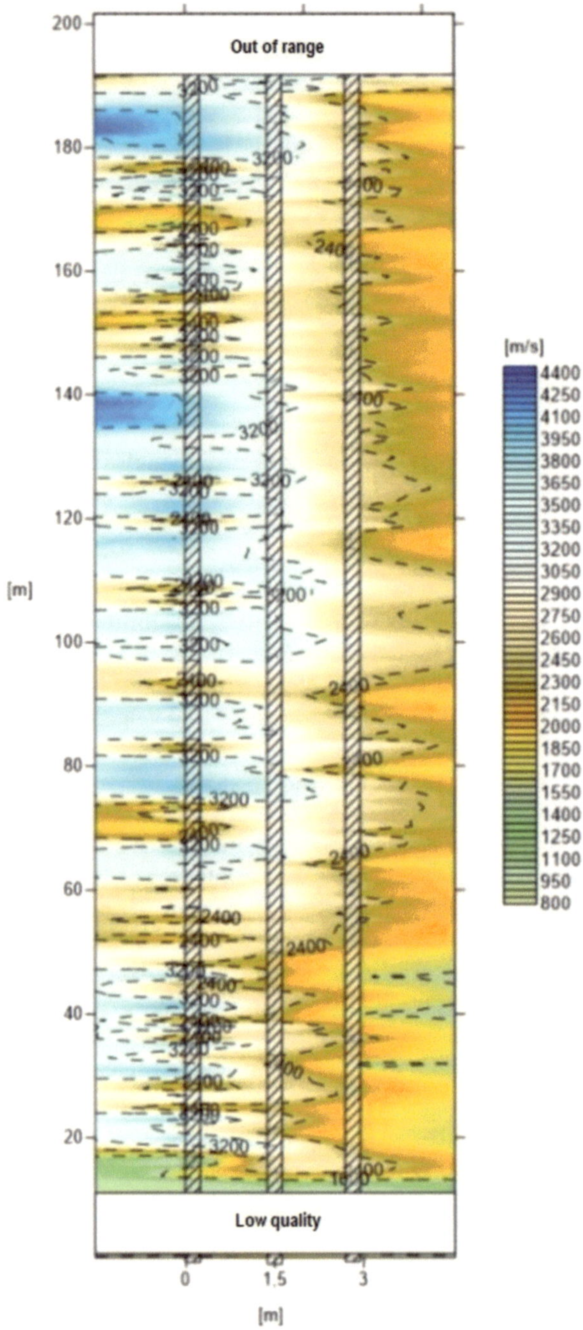

Figure 8.
Map of wave velocity distribution.

1,5 months after test described above, another measurements were carried out, within the stage II of the research. Another map of velocity distribution was made. This map, together with the previous one (**Figure 9**) are presented in the **Figure 13** [18].

On the basis of conducted research, conclusions were made [13, 18]:

- there is a clear relationship between frozen rock mass temperature changes and wave velocity measured. Ground-freezing process monitoring can be conducted using seismic methods.

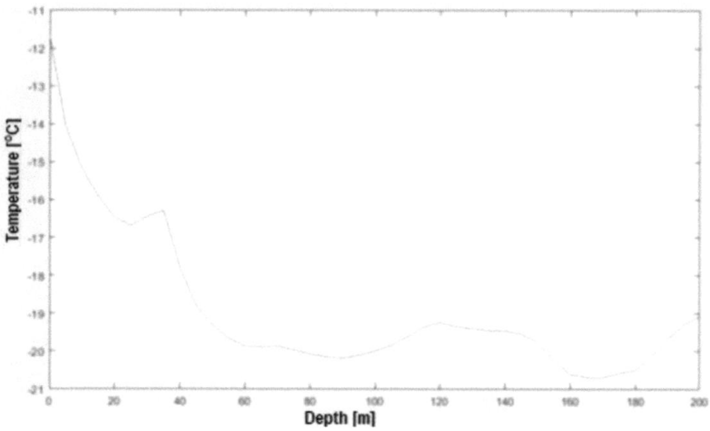

Figure 9.
Temperature in the borehole T1.

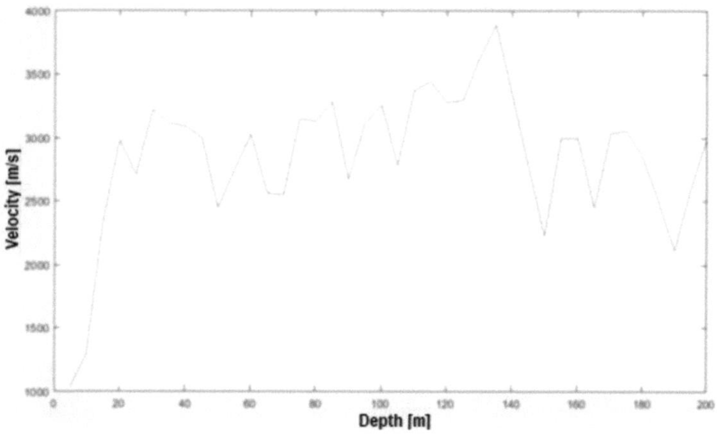

Figure 10.
Velocity in the borehole T1.

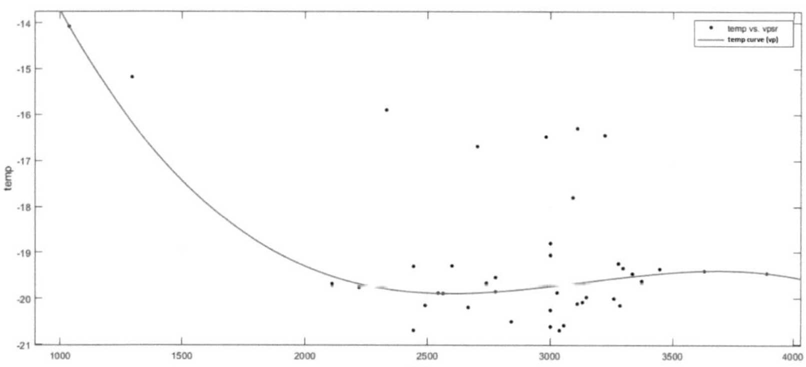

Figure 11.
Relationship between temperature and velocity.

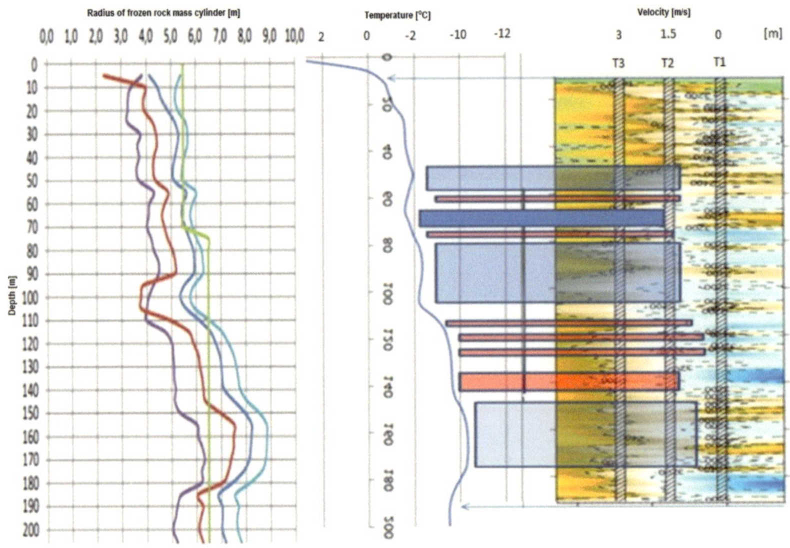

Figure 12.
Graphical correlation between velocity distribution, temperature and frozen rock mass cylinder shape.

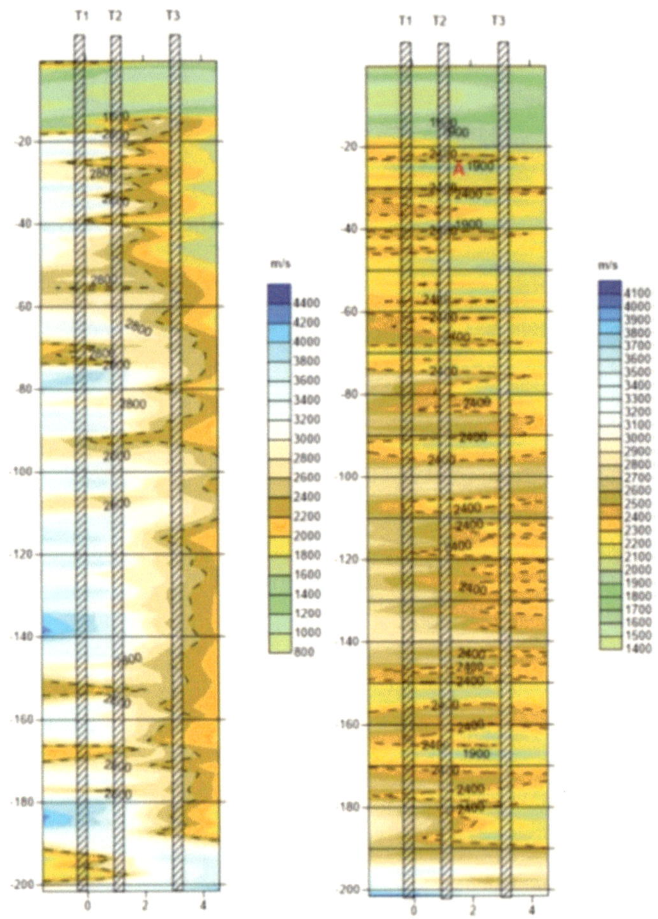

Figure 13.
Maps of velocity distribution.

- development of frozen rock mass column is correlated with velocity. It is in stronger relationship with velocity than just temperature, because temperature value only does not give complete information about geomechanical state of rock mass.

- relative temperature and velocity changes are complementary. Their correlation might have an impact on rock mass imaging state.

6. Summary

Number of seismic methods can be used for measurements of frozen rock mass parameters, such as frozen ground cylinder shape and size during shaft sinking. However, some of them might be too expensive or problematic in use, not effective enough, etc. Important factor is also time needed for measurements.

Different methods were proposed for the purpose of frozen rock mass column monitoring during Grzegorz shaft sinking. Scattered waves interferometry and vertical seismic profiling were assumed the most convenient.

Seismic interferometry method is favorable because of possibility of conducting near real-time monitoring. High sensitivity for temperature changes can provide accurate indication of failures in freezing installation operation. However, technology of such measurements needs further research, spanning for a long time and covering controlled stoppage of freezing installation operation. Financial aspect of this kind of technology is also an issue to consider.

Vertical seismic profiling provides information about shape and size of frozen ground column. Constant measure is impossible, but periodic measurements might find application, because of possibility of determination of shape of frozen rock mass cylinder for tens of meters forward the shaft heading. Therefore, it helps to forecast potential hazards before the beginning of successive sinking stages. Similar to seismic interferometry method, vertical seismic profiling needs further development and analysis of economic factors. However, this method is considered prospective.

To sum up, geophysical methods of frozen ground column monitoring are prospective direction of research. Technological aspects of these methods should be analyzed and developed to make them economically reasonable.

Acknowledgements

This research received no external funding.

Conflict of interest

The author declares no conflict of interests.

Author details

Paweł Kamiński
Faculty of Mining and Geoengineering, AGH University of Science and Technology, Krakow, Poland

*Address all correspondence to: pkamin@agh.edu.pl

References

[1] Jara Ł., Beck M., Drabek J., Kuźma K., Wójcik P., Martyka J., Hajok M., Kamiński P. (2020) *Szyb Grzegorz – raport z budowy*, Inżynieria Górnicza 1/2020, s. 32-37.

[2] Izydorczyk P., Kamiński P., Cichoń P. (2018) *Szyb Grzegorz – raport z budowy. Wieża wyciągowa oraz górnicze wyciągi szybowe,* Inżynieria Górnicza 4/2018, s. 60-62.

[3] Jara Ł., Beck M., Kamiński P. (2018) *Szyb Grzegorz – raport z budowy. Instalacja mrożeniowa*, Inżynieria Górnicza 2-3/2018, s. 47-49.

[4] Kostrz J. (1964) Głębienie szybów metodami specjalnymi, Wydawnictwo „Śląsk", Katowice 1964.

[5] Kostrz J. (2014) *Głębienie szybów*, Biblioteka Szkoły Eksploatacji Podziemnej, Kraków, 2014.

[6] Harris J.S. (1995) *Ground Freezing in Practice*. DOI:10.1016/0148-9062(96)91728-3

[7] Więckowski A., Hajto D., Proficz P. (2018) *Technika zamrażania gruntów i skał luźnych*, Przegląd budowlany, 3/2018, PZITB Warszawa

[8] Alzoubi M.A., Xu M., Hassani F., Poncet S., & Sasmito A. (2020). *Artificial ground freezing: A review of thermal and hydraulic aspects*. Tunnelling and Underground Space Technology, 104, 103534.

[9] Schmall P., Braun B. (2006). *Ground Freezing — A Viable and Versatile Construction Technique*. Proceedings of the International Conference on Cold Regions Engineering. 1-11. 10.1061/40836(210)29.

[10] Sanger, F.J. (1968). *Ground Freezing in Construction*. Journal of the Soil Mechanics and Foundations Division, 94, 131-158.

[11] Williams P. (2005). *Frozen Ground Engineering* (Second edition). Chichester: John Wiley & Sons. xii 363 p, illustrated, hard cover. ISBN 0-471-61549-8. Polar Record, 41(2), 174-175. doi:10.1017/S003224740529441X

[12] Owczarek D., Wójcik P. (2019) *Wiercenie otworów mrożeniowych*, In: Proceedings of the International Mining Forum 2019, 11-12.04.2019, Katowice, Poland

[13] Matuła R., Czarny R. (2019) *Raport. Wyniki zastosowania metod sejsmicznych podczas drążenia szybu Grzegorz. Etap I,* praca niepublikowana.

[14] Sato H., Fehler M.C., Maeda T. (2012) *Seismic Wave Propagation and Scattering in the Heterogeneous Earth*: Second Edition. Vol 9783642230.; 2012.

[15] Snieder R., Gret A., Douma H., Scales J. (2002) *Coda Wave Interferometry for in Seismic Velocity*. Science (80) ;295(March):2253-2255.

[16] Czarny R. (2017). *Ocena budowy i właściwości sprężystych górotworu metodą interferometrii sejsmicznej*, PhD dissertation, IGSMiE, AGH University of Science and Technology, Kraków.

[17] Buske S., Gutjahr S., Sick C. (2009). *Fresnel volume migration of single-component seismic data*. Geophysics, 74(6, S), WCA47-WCA55.

[18] Matuła R. (2019) *Raport. Wykorzystanie metod sejsmicznych podczas drążenia szybu Grzegorz. Etap II*, praca niepublikowana.

Chapter 2

A Novel PID Robotic for Speed Controller Using Optimization Based Tune Technique

Falih Salih Mahdi Alkhafaji, Wan Zuha Wan Hasan, Nasri Sulaiman and Maryam Mohd. Isa

Abstract

One of the most significant issue of proportional integral derivative (PID) controller is the efforts to optimize coefficient gains. Based on survey, massive tuning methods were proposed to resolve this problem but there is little pay attention to maximize minimization time response significantly. This study proposed a novel technique to maximize optimization PID gains for the DC motor controller by combining both proper tuning method with signal input signal output (SISO) optimization toolbox using optimization based tune (*OBT*) techniques, that could be utilized for the highest precision controller. The comparative study has been carried out by applying five different tuning methods to obtain a proper tuning controller, then to be combined with SISO optimization toolbox. The utilized tuning methods are Robust Auto tune (RAT), Ziegler–Nichols (Z-N), Skogestad Internal Model Control (SIMC), Chien Hroues Reswick (CHR), and Approximate M-Constrained Integral Gain Optimization (AMIGO). The performance of each tuning methods based *OBT* are analyzed and compared using MATLAB/SISO tool environment, where the efficiency has been assessed on a basis of time response characteristics (Ti) in terms of dead time (td), rise time (tr), settling time (ts), peak time (tp) and peak overshoot (Pos). The simulation results of AMIGO based proposal show a significant reduction time response characteristic to be measured in the Microsecond unit (μs). The novelty feature of the proposed is that provides superior balancing between robustness and performance. This study has been completely rewritten to account for the robotic controller development that has been taken place in the last years.

Keywords: PID controller, time response, tuning, optimization, AMIGO, DC motor

1. Introduction

A PID controller has been intensively utilized in industries for controlling feedback systems over five decades, in case of simplicity, robustness, flexibility, applicability, and satisfactory performance [1–7]. It became a standard tool and found in many engineering sectors [8, 9], playing an important role in industrial process control. Meanwhile, 60% of loops have bad performance and 25% cannot meet the performance requirements in the industry [10]. Functionally, PID is used to reduce the divergence between the set point and the process variable, which can be employed to upgrade time response and shortening the settling time of a system by tunes properly the gain parameters [11, 12].

Structurally, PID consists of three main proportional parameters, proportional gain (Kp), integral gain (Ki), derivative gain (Kd). The main function of these parameters as follows: Kp is to minimize the tr and steady-state error (Ess); Ki is to eliminate the steady-state error; Kd is to augment the system's stability, minimizing the overshoot, and enhancing the time response specification. The transfer function (TF) of PID might be realized in z-domain and s-domain [13–16]. Several criteria that influence the performance of the controller; type of algorithm, the efficiency of the tuning method, and the complexity of the design. The significant issue in the PID controller is how to tune proportional gains properly [17]. Various tuning methods had been implemented for the tuning PID controller to improve the time response specifications of the plant system by adjusting the three proportional gains. The time response specifications could be assessed in terms of tr, ts, PoS, and Ess [18]. For two decades, the process control industry has seen numerous advances as far as tuning techniques and controller design [13]. Although the PID controller has just three proportional gains, but it is quite strenuous to find out the best gains to meet desired adjustments. Enhancing the PID controller's performance might be accomplished when taking in consideration a systematic procedure of tuning proportional gains, otherwise, it is poorly tuned and raises the consumed time through process tests [3, 19]. PID tuning is extremely used to improve controller performance such as short transient, high stability makes this process harder. Practically, tuning PID gains appear to be impulsive and troublesome. The proper controller should be able to provide a system better stability by eliminating oscillation in any condition of set point [9, 20]. As shown in **Figure 1**, the tuning controller considered a vital branch of control engineers which occupied the majority of hits on the website [21]. Massive various tuning methods had been proposed to achieve satisfying control design in terms of time response specifications tr, ts, PoS, Ess. Some of these methods considered just a single of these objectives as a criterion for their tuning methods, while others considering more than one. The early tuning studies focusing on classical methods such as the Z–N oscillation method, Z- N reaction curve, Cohen Coon curve, and CHR. These traditional methods are extensively applied in cases of ease to utilize [6, 11, 22, 23]. **Figure 2** shows the comparative performance between different classical tuning methods that were used to tune the PID controller based second order system, see more details in [24]. Essentially, tuning controller methods are classified into two main sorts: open loop which indicates to tune the controller when it is in a manual state and the plant operates in an open loop and closed loop which alludes tune the controller during an automatic state in which the plant is known to process and operating in a closed loop [20]. In contrast, conventional tuning methods are still extremely utilized in an industrial

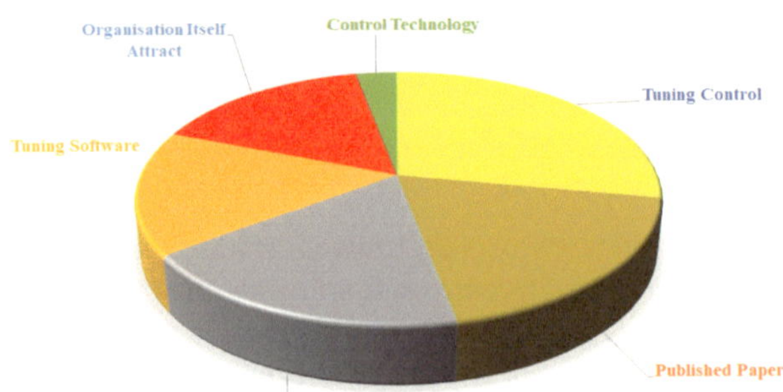

Figure 1.
A survey of PID tuning hints in website.

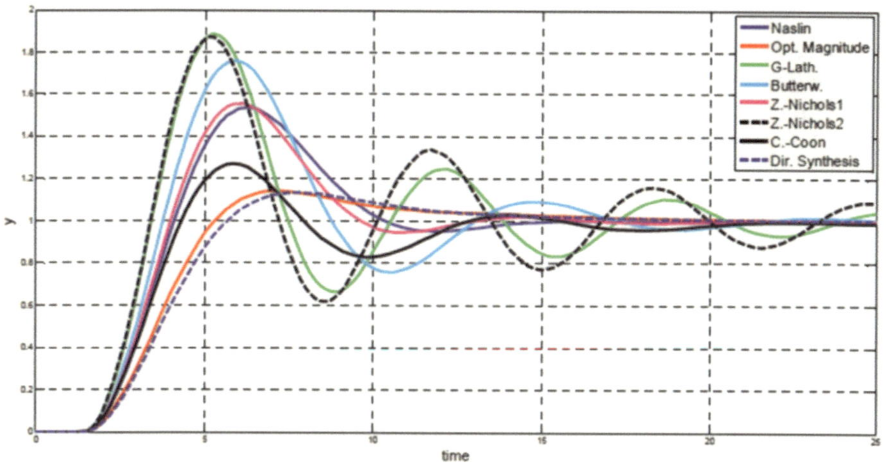

Figure 2.
Time response specifications of PID controller under different classical tuning methods [24].

controller, but they are insufficient to obtain optimal system responses besides, require additional modifications [23–26].

These disadvantages come from two reasons: 1) lacking procedure data and performance leads to poor damping; 2) the performance of the controller is affected by the parameter's variation in dynamic systems. Unstable processes are quite harsh to control compared with stable processes, and the settling time and overshoot are relatively larger for unstable systems than that of stable systems. However, development tuning methods are difficult and impractical purpose. Further, some methods are inapplicable for higher ratios of a time delay to time constant such as greater than 0.9 [13, 26]. The better tuning method depends on several criteria as follows: 1) the tuning rules should be analytical; 2) more simplicity and convenience to memorize; 3) applicable to be used in a wide range of processes [19, 27]. The majority of problems in dynamic systems are that the difficulties of describing the real plant exactly, in case of unbalancing behavior between the controller and plant system. It is necessary to extend the abilities of PID controllers to contain extra features. As some methods are better than others for various applications, each one has a negative and positive side. Several studies concentrated on the drawbacks to developing tuning methods by reducing the complexity of design and time of implementation [28–30]. The classification of the Robustness PID controller relies on what extent the robustness to minimize responses, repetition, and avoiding the growing delay time during the calculation process [19]. By contrast, optimization algorithms are effectively being implemented for adjusting gains and minimizing time response. Conventional optimization algorithms are relying on several assumptions such as differentiability, the convexity of the cost function and constraints that should be fulfilled. Optimization algorithms give better results after every iteration [31–33]. Genetic Algorithm (GA) one of the famous optimization algorithms that was invented through the 70s of the last century, which depends on parallel search techniques to adjust PID gain. However, GA suffers from the massive computational through the optimization process [4].

On the other side, a few tuning methods are extremely utilized, for example, Internal Model Control (IMC), which uses to diminish the error by expecting the output, further modifying the proportional gains to achieve the desired set-point response with accepted overshoot [34, 35]. By contrast, there are tremendous tuning methods that have been implemented to improve the precision of a controller system, but less attention has been paid in the review to combine sufficient tuning with an optimization algorithm to maximize optimization with a reduction time response. The survey shows relatively lesser studies focusing on the comparative tuning methods and limited studies for unstable systems. Additionally, there are several limitations in previous works that comes from less paid attention to take in consideration the better balancing between the overshoot and the time response parameters through tuning gains [13, 36].

In this study, we proposed a novel methodology by combined proper tuning with signal input signal output (SISO) optimization technique for adjusting PID gains to maximize minimization time response specifications of DC motor to become more proficient speed. The proposed methodology relies on a comparative study of using five different tuning methods: RAT, Z-N, CHR, Cohen, and Coon method, SIMC, and AMIGO method, applied to second order system. Afterward, proceeding a comparison between tested tuning to verify a proper tuning to be combined with optimization SISOtool. Practically, it was analyzed the time response specifications to those tested tuning methods separately and jointly with the proposed technique to estimate them in terms of td, tr, ts, and PoS.

This paper has been divided into six sections. Section 2 describes the tuning methods under test. Section 3 explained the proposed methodology. Section 4 presents simulation results based on tuning methods and proposed technique. Section 5 discusses the comparison results and PID formulation. Ultimately, the conclusion is summarized in Section 6.

2. Tuning methods under study

The problem of identification PID gains appeared when parameters of a current controller have to be tuned. In simply, if controller parameters are tuned manually with time, the controller does not achieve satisfactory performance. One important factor to be considered in designing a controller relatively with a plant is the efficiency of tuning techniques [9]. In this study, we used five different tuning methods tested with specific second order models to make a comparison and to verified the best method, prepared to combine with optimization SISO tool. In this section, we give a brief description of the selected tuning methods.

2.1 RAT tuning application

The Robust Control toolbox permits to tune control systems for robustness against parameter variation of the process, and to ensure performance across a range of operating conditions. Further, it can be used for multi model tuning to specify the best controller parameters for all plant models. There are several approaches that can be used in this application to improve tuning depends summarizes as follows: 1) Tune control system for robustness against parameter uncertainty; 2) tune fixed-structure control system against real and complex parameter uncertainty and dynamic uncertainty; 3) tune controller for a few critical values of the process parameters; 4) ensure performance across different operating conditions; 5) tune for satisfied controller over multiple system configurations [37].

2.2 Z-N tune

Z-N is a famous tuning method based on closed-loop experiments, which is widely using in process control, to achieve stability for a plant system of which the mathematical models are unknown or difficult to obtain [3, 7, 20]. It is classified into two sets, step response method and frequency response method. Z-N relies on two objectives: 1) To characterize process dynamics by two parameters, that are specified experimentally; 2) using a simple formula to figure out controller parameters from the time response characteristics. The Z-N method has had a very strong impact on the practice of the controller [1, 38]. Theoretically, Z-N relies on defining the value of Kp, Ti and Td cantered on the time response of the plant. If the model system does not have integrator and dominant merge poles, then the S-shaped curve will be shown by the unit step response curve. **Figure 3** shows the curve has two constants, time constant (T) and delay time (L) that can be obtained by drawing a tangent at the inflection point of the curve and locate the intersection of the tangent with the time axis and line c (t) = K. The compensator TF of PID controller can be obtained by setting Kp = 1.21/a, integral time Ti = 2 L and derivative time TD = 0.5 L, where a = (K.L)/T [3]. By contrast, Z-N has several drawbacks, for instance, once the controller is tuned by the Z-N method, good adjusting but not accomplished optimal responses, leads to face a lot of obstacles through implementation. Further, it suffers from time consuming, which may require many trials to obtain optimal gains, and inapplicable for unstable open loop system. Moreover, it is suitable to be used with a specific plant, but the transient response can be even worse when the system fluctuates. Additionally, due to a set point's variation or external disturbances, this leads to forces the process into an unstable operation situation. Fixing PID parameters causes leakage responses and bad performance indices. Consequently, Z-N tuning without modifications does not work well with all processes. However, these disadvantages can be resolved by using a simple modification for processes to overcome the leakage of the time delay [6, 18, 20, 39].

2.3 SIMC tune

To overcome the disadvantages over classical methods, a new merged structure with PID was proposed IMC- PID relay on pole-zero conversion for stable and unstable processes with time delay. In 2010 Shamsuzzoha and Skogestad proposed a modification to the Z-N symbolized SMIC referred as Internal Model Control that could be used to enhance the PI/PID controller for an unidentified process by utilizing closed-loop experiments. This method relies on classical ideas presented earlier by Ziegler and Nichols. The importance of this method is that there is a single tuning parameter was proposed to modify PID gains optimally and to obtain better

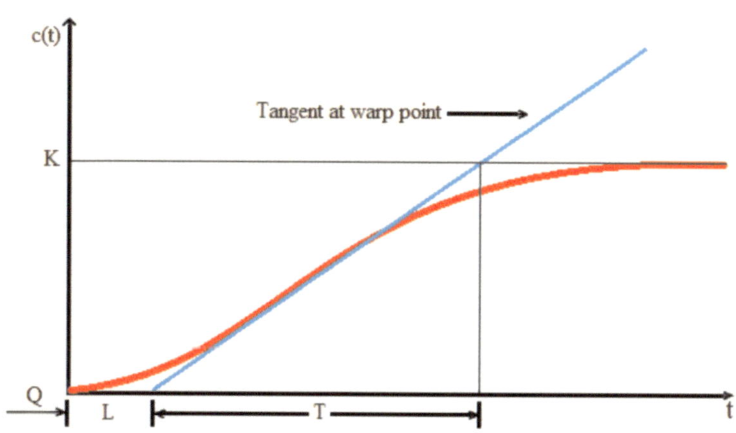

Figure 3.
Step response of plant based Z-N tune [3].

balancing between performance and robustness without needing any experimental tune, but this method still needs additional modification to get optimal gains [6, 12].

An IMC-PID controller obtains superior performance and adequate set-point tracking but displays low responses to disturbances, especially for lag-dominant including integrating processes. This approach has several advantages, such as considers model uncertainty and allows the designer to trade-off control system performance against control system robustness, to process changes and modeling errors. It is well known that the tuning parameters should be selected in a moderate way to achieve justified balancing of both robustness as well as performance. The accuracy of IMC tuning can be determined by the effective influence on the performance of the controller, which fundamentally relies on the construction of the IMC filter. Consequently, the ideal IMC filter structure must be chosen considering the performance of the resulting PID controller rather than that of the IMC controller. In this method, a lesser value for tuning parameters gives a good nominal performance, where a higher value gives robust control performance with a compromise on nominal responses. There is another drawback with traditional IMC based PID approaches is that the IMC filter is usually selected based on the performance of the resulting IMC tune not on the time response specifications of the PID controller. For this reason, it is very beneficial to convert an IMC controller to a PID controller (IMC-PID approaches) to increase the performance reduction responses. Significantly, the resulting PID controller could have poor control performance when an IMC filter structure involves an error in its conversion to the PID controller, despite being derived from the best IMC performance. The accuracy of PID based IMC tune relies on both dead time approximation error and the conversion error that relates to the filter structure and the process model. Hence, to improve PID performance, it should be taken into consideration the best roles to design the IMC filter optimally for each specific process model. As shown in **Figure 4**, the IMC structure constructed from three block: 1) Gp is referred to process; 2) Gm is referred as the process model; 3) GcI is the IMC controller. This method contributes to reduced integral time for processes, but with a large process time constant. Theoretically, this system relies on extra new parameters that can be used significantly to minimize responses by optimizing PID gains. **Figure 5** shows these parameters: Controller gain (Kc0); set point change (Δys); time from set point change to reach a first maximum peak (tp); corresponding maximum output change (Δyp); output change at first undershoot (Δyu) [13, 20, 39].

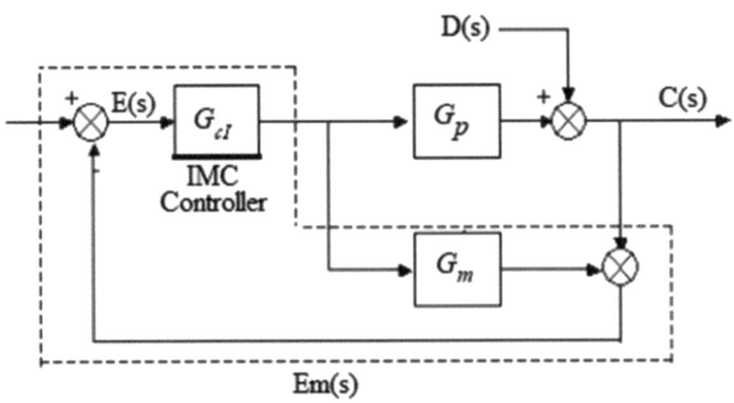

Figure 4.
IMC structure [6].

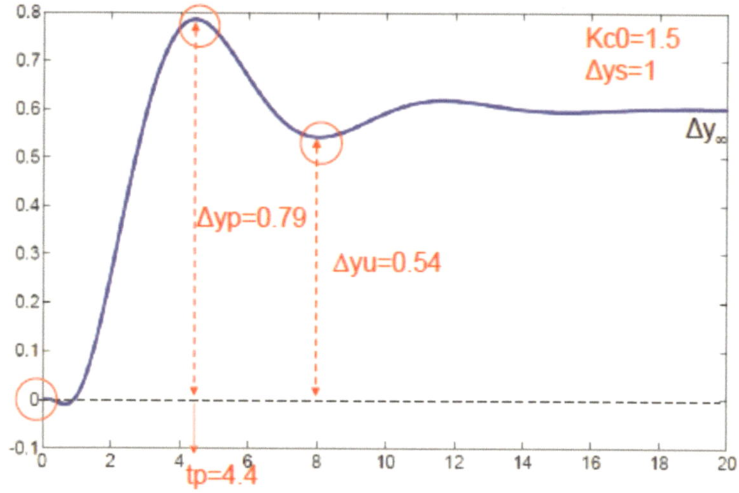

Figure 5.
Extracting parameters from IMC closed-loop set point response with PID controller [20].

2.4 CHR tune

In 1952 Hrones and Reswich proposed a modification of the open loop Z-N method to improve the time response and overshoot. They proposed to use the fastest responses with a 20 percentage overshoot as a design criterion. This method can be used for point regulation or noise rejection to control the responses to obtain better performance even at higher order system [3, 6, 40]. The set point tracking of PID controller parameters can be specified for zero percentage overshoot Kp = 0.6/a, Ti = T, TD = 0.5 L, where the parameters value in 20 percentage of the overshoot are Kp = 0.95/a, Ti = 1.4 T, TD = 0.47 L. Tuning rules based on 20 percentage overshoot design criteria are quite like the Z-N method. On the other side, when zero percentage overshoot criteria are utilized, the integral time is larger and the gain and the derivative time are smaller, that's mean a proportional action and an integral action, as well as a derivative action, are smaller. By contrast, both of Z-N and CHR method use the time constant T delay time L and gain k of the plant explicitly [1, 6, 8].

2.5 AMIGO tune

The procedure of the AMIGO tune is similar to that of the Z-N tune method. This method gives a controller more capability to reduce load disturbances effectively, besides maintaining good robustness. Theoretically, the time constant of this method can be obtained by Aström's equations. Essentially, the design procedures performed on loop shaping provides satisfaction integral gain subject to a robustness constraint. The robustness could be verified in terms of the maximum value of the sensitivity function. Therefore, the AMIGO tune can be represented as a minimization of the integrated error at step load disturbances subject to constraints on the maximum sensitivity [3, 38, 41].

3. Methodology

This section describes the system under study and proposed technique of how to use both tuning PID with optimization SISO tool to minimize Ti and Pos significantly. The novelty of the proposed technique is to maximize the minimization time response of the controller as will be explained in the following sections.

3.1 Mathematical model of speed DC motor

To design a PID controller precisely, it is essential to evaluate the mathematical model of the selected DC motor, then applying the proposed methodology. The mathematical model of the DC motor was derived based on Eq. (6). In this model we assume that the back emf (Ke) and motor torque (Kt) are equal, therefore:

$$Kt = Ke = K \tag{1}$$

$$Ki = J.\ddot{\theta} + \mathrm{b}.\dot{\theta} \tag{2}$$

$$V - K\dot{\theta} = L.\frac{di}{dt} + Ri \tag{3}$$

By applying the Laplace transform to get s-domain equations:

$$s(Js + b)(s) = KI(s) \tag{4}$$

$$(Ls + R)Is = V(s) - Ks(s) \tag{5}$$

Where, L is the electric inductance, R is the resistance of the coil, b is motor viscous friction constant, $\theta$is the speed of the shaft, Kt motor torque constant, Ke electromotive force constant, and J is the moment of inertia of the rotor.

By eliminating I (s), it can be derived the mathematical model of speed DC-Motor, and substituted the following DC motor parameters in Eq. (6). The speed mathematical model of the selected DC motor can be derived as given in Eq. (7):

L = 0.5H, R = 1 Ω, b = 0.1 N.m. s, Kt = 0. 01 N.m/Amp, Ke = 0.01 V.sec/(rad), J = 0.01 kg. m^2.

$$P(s) = \frac{\dot{\theta}}{V(s)} = \frac{K}{(Js + b)(Ls + R) + K^2} \frac{[rad/sec]}{V} \tag{6}$$

$$P(s) = \frac{0.01}{0.005\,S^2 + 0.06\,S + 0.1001} \tag{7}$$

3.2 Proposed technique

It was suggested a new technique to optimize the PID gains by following two stages. Initially, we utilized five different traditional tuning methods RAT, Z-N, SIMC, CHR, and AMIGO to tune the PID controller lonely, and to produce compensator TF prepared to be imported to the next stage. Secondly, to improve the reduction Ti and Pos for the chosen model by using *OBT* technique, where this methodology needs a benchmark comparison to nominate the best tuning method to be joined with the SISO toolbox application. In view of the two ideas suitability of optimization and time response parameters, we perform a comparison between tuning understudy independently and jointly to prove the validation of the proposed technique, using the SISO optimization based tune ($\boldsymbol{SISO_{OBT}}$). All tuning methods under study are mimicked by giving a unit step change in set point.

The $\boldsymbol{SISO_{OBT}}$ technique provides two significant aspects. Firstly, how changes in the TF's poles and zeros alter the root locus and bode plots in live response, besides tracking the changes in gains that can influence to system's step response. Secondly, specifying the design requirements will create visual limitations on the graphs to help the user set and find appropriate gains. The time response of the DC motor model was simulated using MATLAB code. The Ti results based proposed technique have been contrasted with each other existing tuning methods. This is done by chosen" Design Requirements" on the SISO tool root locus plot set and define its limits on the characteristic of the time response. It was investigated the criteria which affect the controller performance in terms of td, tr, ts, steady state time (SST), tp, Overshoot PoS, and number of Iterations. **Figure 6** presents the proposed technique to optimize PID gains which rely on several steps as follows: 1) Import the mathematical model into estimation tool; 2) putting compensator gain to 1 and selecting r to y for negative feedback controller using step response for analysis plot; 3) measuring the time response specifications by LTI viewer; 4) adding the PID controller to a system and applied the tested tune based step response to extract proportional gains, then updating the compensators gain and TF. This is the first stage of adjusting PID gains; 5) preparing to optimization stage, on LTI viewer specifying the design requirements as the following; tr 1e-8 for 90%, ts 2e-8, *PoS* 2; 6) readjusting the location of compensator Zeros for fine modification proportional gains, using optimization SISO closed loop r to y; 7) from LTI viewer import TI and Pos, it can be seen that there is a magnificent reduction time response contrasted with the first level tuning; 8) import the TF of compensator with optimization proportional gains; 9) using switching mod algorithm to select one by one different tuning methods RAT, Z-N, SIMC, CHR, AMIGO and repeat those steps 1–8 for each one. The last step provides comparison results between tuning under study based proposed methodology to select a suitable tune that could be chosen to finalize the design based proposed methodology.

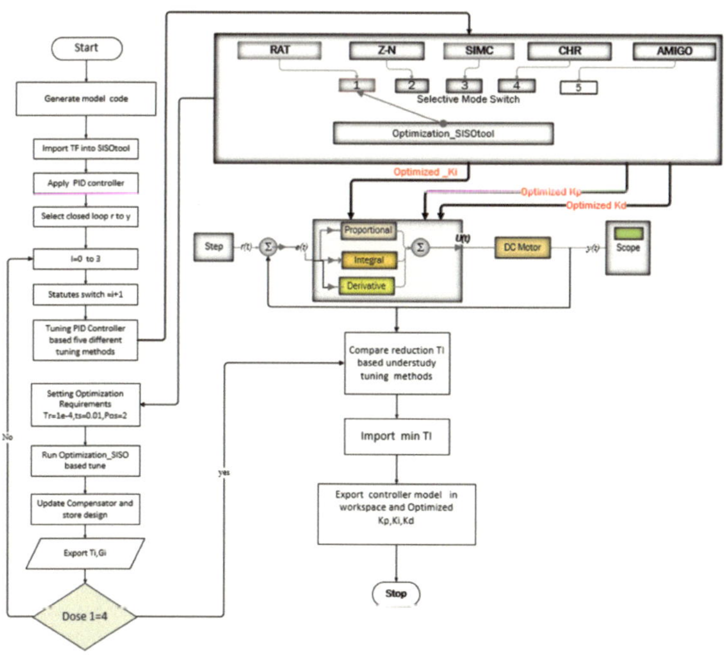

Figure 6.
The PID controller based proposed technique.

4. Results and analysis

This section illustrates the simulation results in terms of time response specifications, proportional gains and poles zeros locations. The performance of the proposed method was tested with the TF of the plant model as given in the aforementioned Eq. (7). **Figure 7** shows the transient response of an uncontrolled system under study. It can be seen that the system is very slow response measured in second (s) unit, where the time response characteristics are: td = 44 ms, tr = 0.627 s, ts = 1.13 sec, tp1.85 s, SSt 1.8 s, and peak amplitude 0.0476. The proposal relies on two levels of tuning. Firstly, we used RAT, Z-N, SIMC, CHR, and AMIGO tuning methods to tune the PID controller separately and to obtain a comparison between them in terms of time response characteristics. Further to generate a compensator TF prepared to be used as a modified system. The obtained gains are used as a first tuning level. Secondly, combine tuning methods with optimization SISO toolbox application using *OBT* technique to optimize PID gains generated by the first tuning level. Where the obtained results are used as a second tuning level.

Figures 8–12 presents the Ti characteristics of the system under separately tuning and based *OBT* for the tested tuning methods. **Figure 8(a)** shows the time response of the system using the RAT lonely. The obtained results are: PoS = 6.24%, td = 0.102 s, tr = 0.267 s, ts = 0.935 s, constant time = 1.5 s, where tuning gain values are: Kp = 319, Ki = 1/Ti = 1.79e+03 and Kd = 4.22. **Figure 8(b)** shows the time response by RAT based *OBT* (**RAT_{OBT}**). It is obviously seen that there is significantly improved time response and overshoot as follow: td = 0.0943 ms, tr = 0.31 m s,

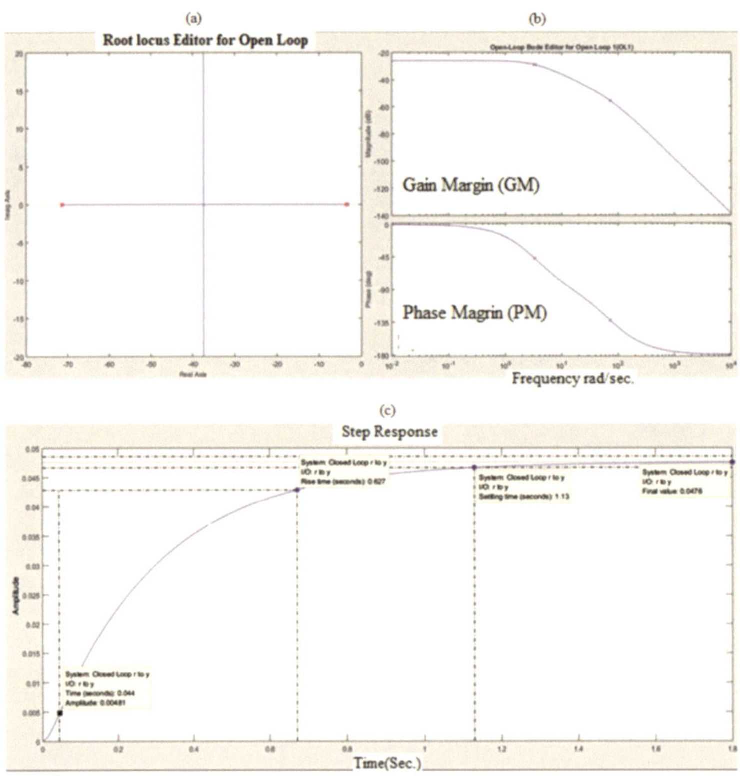

Figure 7.
Tested uncontrolled system, (a) poles zeros location, (b) gain and phase margin, (c) time response characteristics.

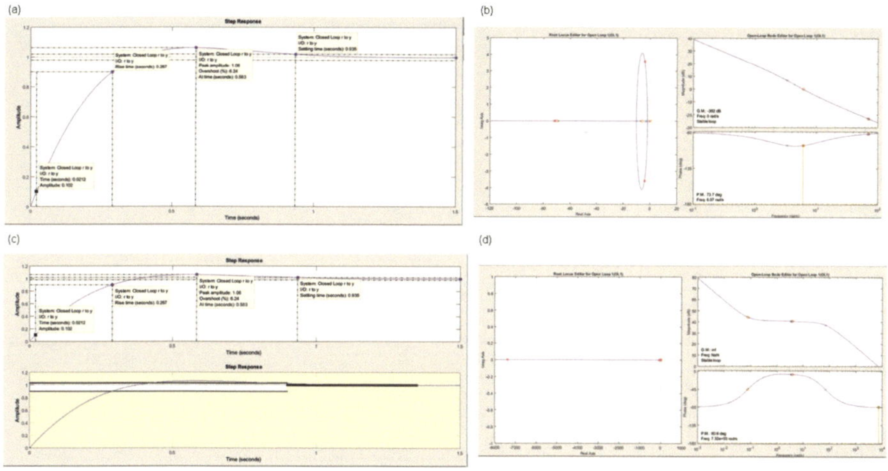

Figure 8.
Step response and poles zeros location of controller system based RAT separately tune and RAT_{OBT}, *(a) step response based RAT separately tune, (b) poles zeros location and gain/phase margin based separately tune, (c) step response based* RAT_{OBT}, *(d) poles zeros location and gain/phase margin based* RAT_{OBT}.

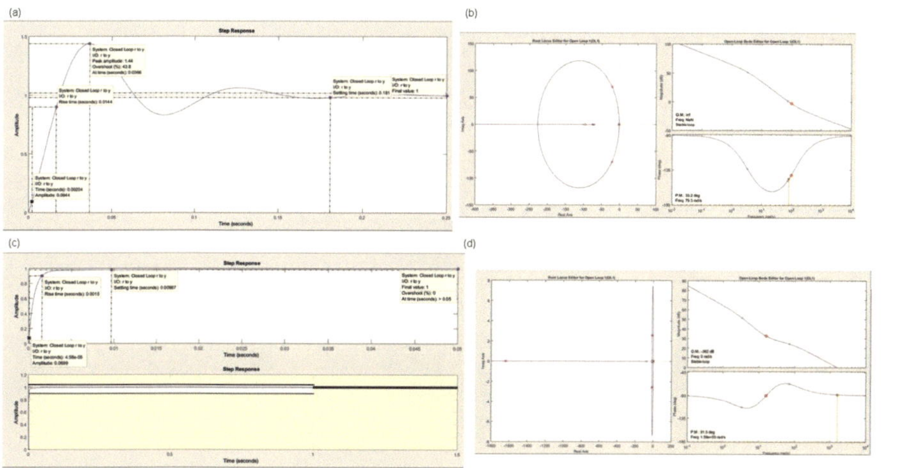

Figure 9.
Step response and poles zeros location based Z-N tune and Z_N_{OBT}, *(a) step response based Z-N separately tine, (b) poles zeros location and gain /phase margin based Z-N separately tune, (c) step response based* Z_N_{OBT}, *(d) poles zeros location and gain /phase margin based* Z_N_{OBT}.

ts =0.617 ms, constant time = 1.4 ms, PoS = 0%, where the tuning gains values are; Kp = 4.6888, Ki = 1/Ti = 181 and Kd = 615. **Figure 8(c,d)** show the root locus of controller based RAT lonely and $\boldsymbol{RAT_{OBT}}$ respectively,it is obviously to seen that the modification of zeros location based $\boldsymbol{RAT_{OBT}}$ produces high adjusting location compared with other tuning based *OBT*. **Figure 9(a)** shows the time response of the model based Z-N tuning lonely. The obtained time response results are:
PoS = 43.8%, td = 0.0944 s, tr = 0.0144 s, ts = 0.181 s, constant time = 0.25 s, where tuning gain values are: Kp = 13.9269, Ki = 1/Ti = 10.656875 and Kd = 36.2. **Figure 9(b)** shows the time response produced by Z-N based *OBT* ($\boldsymbol{Z\_N_{OBT}}$). It is obviously seen that there is significantly improved time response and overshoot as follow: td = 2.2 Micro s, tr = 0.0015 s, ts = 0.00967 s, T = 0.05 s, PoS = 0%., where the tuning gains values are: Kp = 10.67551, Ki = 1/Ti = 10.6568757 and Kd = 2.5938410. **Figure 9(c,d)** illustrate the root locus of the controller based Z-N lonely and $\boldsymbol{Z\_N_{OBT}}$ respectively.

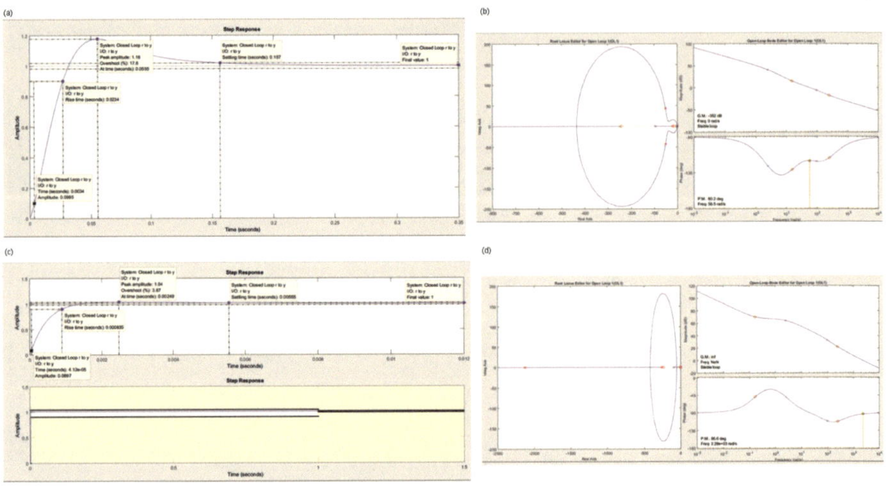

Figure 10.
Step response and poles zeros location based SIMC tune and $SIMC_{OBT}$, *(a) step response based SIMC separately tine, (b) poles zeros location and gain/phase margin based SIMC separately tune, (c) step response based* $SIMC_{OBT}$, *(d) poles zeros location and gain/phase margin based* $SIMC_{OBT}$.

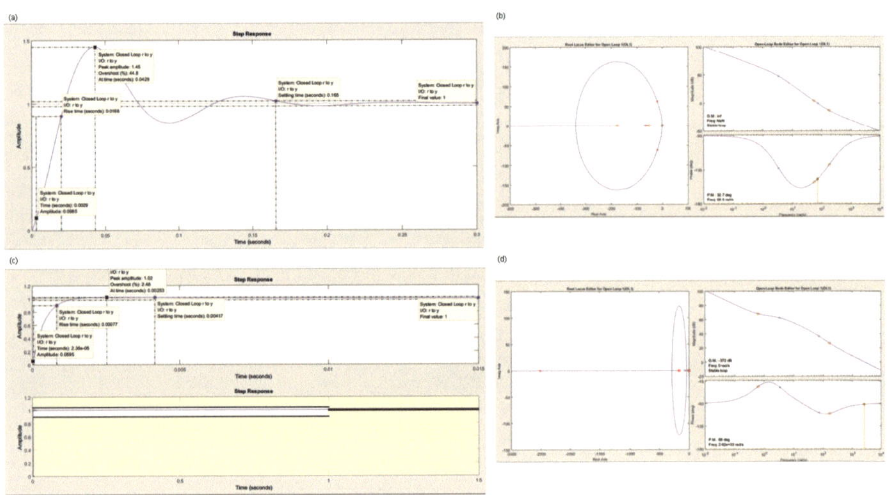

Figure 11.
Step response and poles zeros location based CHR tune and CHR_{OBT}, *(a) step response based CHR separately tine, (b) poles zeros location and gain/phase margin based CHR separately tune, (c) step response based* CHR_{OBT}, *(d) poles zeros location and gain/phase margin based* CHR_{OBT} .

It is obviously seen that the modification of zeros location based $\boldsymbol{Z\_N_{OBT}}$ produces high adjusting location compared with tuning lonely.

Afterward, we will investigate the tuning gains based SIMC separately and jointly with optimization SISO application, and what happens to time response characteristics with overshoot. **Figure 10(a)** presents the time response of the model based SIMC tuning method. The obtained results are: td = 0.0034 s, tr = 0.0234 s, ts = 0.157 s, constant time = 0.35 s, Pos = 17.8%, where tuning gain values are; Kp = 5.29e+0.3, Ki = 1/Ti = 7.55e+0.4 and Kd = 20.6. **Figure 10(b)** shows the time response by SIMC based proposed method ($\boldsymbol{SIMC_{OBT}}$). It is obviously seen that there is significantly improved time response and overshoot as follow: td = 4.12e-05 s, tr = 0.835 ms, ts = 5.5 m s, constant time = 0.012 s, PoS = 3.87, where the tuning gains values are: Kp = 4.54e+05, Ki = 1/Ti = 7.55e+04 and Kd = 1.92e+03.

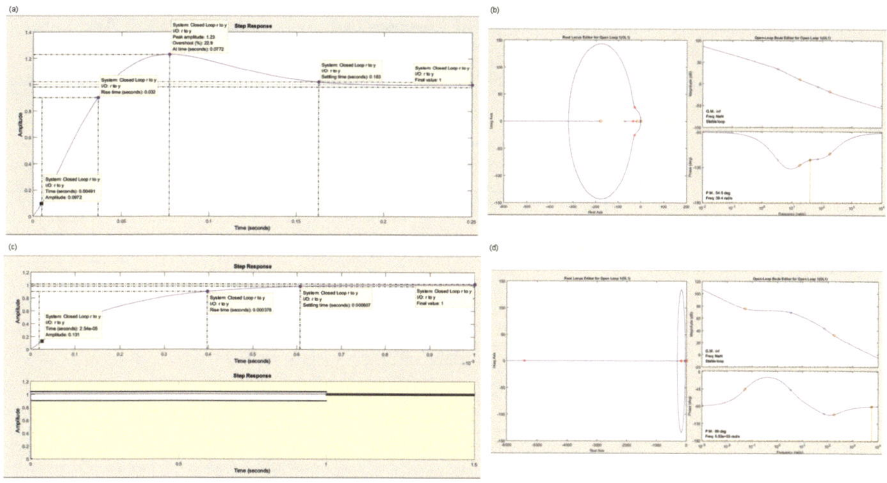

Figure 12.
Step response and poles zeros location based AMIGO tune and $\text{AMIGO}_{\text{OBT}}$, *(a) step response based AMIGO separately tine, (b) poles zeros location and gain/phase margin based AMIGO separately tune, (c) step response based* $\text{AMIGO}_{\text{OBT}}$, *(d) poles zeros location and gain/phase margin based* $\text{AMIGO}_{\text{OBT}}$.

Figure 10(c,d) illustrate the root locus of the controller based SIMC lonely and $\boldsymbol{SIMC_{OBT}}$ respectively, it is obviously to seen that the modification of zeros location by $\boldsymbol{SIMC_{OBT}}$ produces high adjusting location compared with other tuning based *OBT*. **Figure 11(a)** shows the time response of the system using CHR tuning method lonely. The obtained results are: td = 0.0029 s, tr = 0.0168 s, ts = 0.165 s, constant time = 0.3 s, PoS = 44.8%. Tuned gain values are: Kp = 553, Ki = 1/Ti = 2.24e+04 and Kd = 2.41. **Figure 11(b)** illustrates the time response produced by CHR based *OBT* ($\boldsymbol{CHR_{OBT}}$) as follow: PoS = 2.48, td = 2.36e-05 s, tr = 0. 77 ms, ts = 4.1 m s, constant time = 0.015 s, where the tuning gains values are: Kp = 3.62e+04, Ki = 1/Ti = 2.24e+04 and Kd = 219. **Figure 11(c,d)** illustrate the root locus of controller based CHR lonely and $\boldsymbol{CHR_{OBT}}$ respectively,it is obviously to seen that the modification of zeros location by $\boldsymbol{CHR_{OBT}}$ produces high adjusting location compared with tuning lonely.

Figure 12(a) shows the time response of the system using AMIGO tuning method. The obtained simulation results are: PoS = 22.9%, td = 0.00491 s, tr = 0.032 s, ts = 0.1635s, constant time = 0.25 s, where, tuning gain values are: Kp = 2.66e+03, Ki = 1/Ti = 4.24e+04 and Kd = 13.6. **Figure 12(b)** shows the time response by AMIGO based proposed method ($\boldsymbol{AMIGO_{OBT}}$). It is obviously seen, that there is significantly improved in time response and overshoot compared with others as the following results: Pos = 0%, td = 2.54e-05 s, tr = 0. 378 ms, ts = 0.607 m s, constant time = 1 ms, where the tuning gains values are: Kp = 8.18e+03, Ki = 1/Ti = 4.24e+04 and Kd = 4.64e+03. **Figure 12(c,d)** present the root locus of the controller based AMIGO lonely and $\boldsymbol{AMIGO_{OBT}}$. It is clearly seen that the modification of zeros location based $\boldsymbol{AMIGO_{OBT}}$ produces a high adjusting location compared with compared with tuning lonely.

Table 1 shows the TF and proportional gains of the compensator based *OBT* for different tuning methods. **Figure 13** shows the modification location poles (P) zeros (Z) and adjusting compensator gains, to presents the effectiveness of using $\boldsymbol{AMIGO_{OBT}}$ over another tuning understudy. **Figure 13(a)** presents the values of P and Z for optimization under $\boldsymbol{RAT_{OBT}}$ as follows: 0, Z1 = −0.0785897621, Z2 = -3.7453341264 respectively, and the gain (G) = 614.6118. **Figure 13(b)** shows the values of P and Z produces by $\boldsymbol{Z\_N_{OBT}}$ as follows: P = 0,

	Method	TF of Compensator	Gain	Peak Amplitude
Classical Tuning	RAT	$\frac{4.2238(s+69.52)(s+6.092)}{s}$	4.22	1
	Z-N	$\frac{36.173(s+96.5)^2}{s}$	36.30	1
	SIMC	$\frac{20.561(s+242.3)(s+15.15)}{s}$	20.56	1
	CHR	$\frac{2.4055(s+177.2)(s+52.58)}{s}$	2.4	1
	AMIGO	$\frac{13.632(s+177.5)(s+17.52)}{s}$	13.63	1
Proposed Technique	$\boldsymbol{RAT_{OBT}}$	$\frac{614.61(s+0.07859)(s+3.745)}{s}$	614.61	0.999
	$\boldsymbol{Z\_N_{OBT}}$	$\frac{1340.3(s+15.85)^2}{s}$	1340.25	1
	$\boldsymbol{SIMC_{OBT}}$	$\frac{1915.4(s+236.9)(s+0.1663)}{s}$	1915.36	1
	$\boldsymbol{CHR_{OBT}}$	$\frac{219.34(s+164.4)(s+0.6216)}{s}$	219.34	1
	$\boldsymbol{AMIGO_{OBT}}$	$\frac{4644.6(s+176.1)(s+0.05183)}{s}$	4644.6	1

Table 1.
Comparison modeling compensator between tuning methods under study and proposed technique.

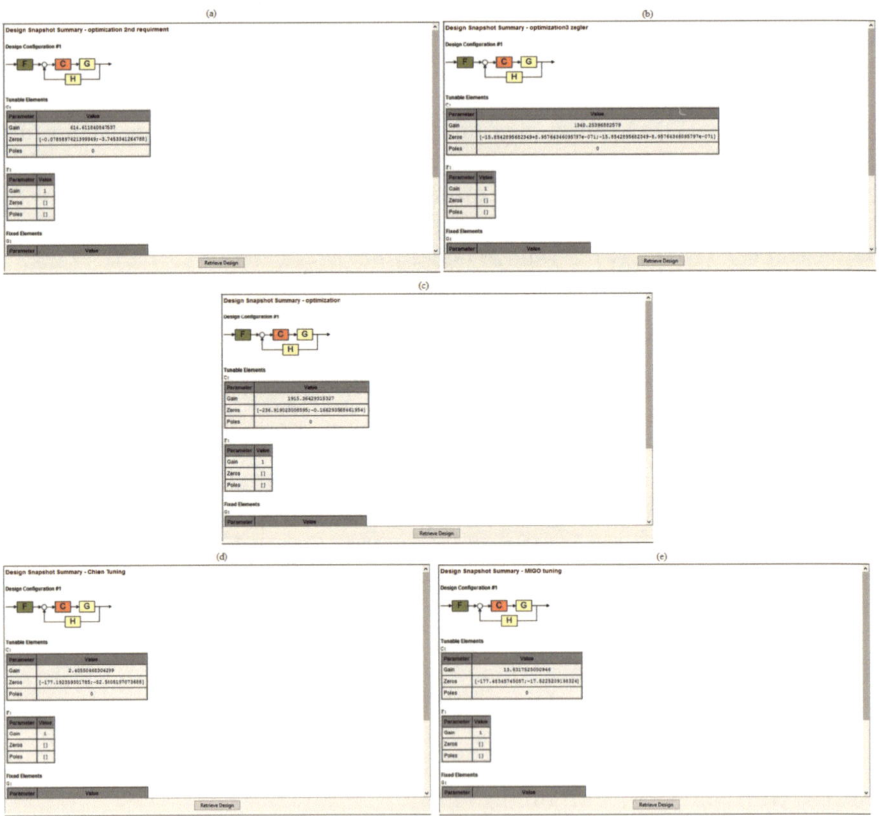

Figure 13.
Optimization results under various tuning methods in term of P, Z, Gains. (a) optimization under RAT_{OBT}*, (b) optimization under* Z_N_{OBT}*, (c) optimization under* $SIMC_{OBT}$*, (d) optimization under* CHR_{OBT}*, (e) optimization under* $AMIGO_{OBT}$.

Z1 = −15.854289 + 8.9576434e-07i, Z2 = -15.854289–8.9576434e-07i, and G = 1340.2539. **Figure 13(c)** shows the values of P and Z on optimization under $\boldsymbol{SIMC_{OBT}}$ as follows: 0, Z1 = −236.91902, Z2 = -0.1662935 respectively, and

G = 1915.3642. **Figure 13(d)** shows the values of P and Z for optimization under CHR_{OBT} as follows: 0, Z1 = -177.192359, Z2 = -52.580519, and G = 2.405504. **Figure 13(e)** shows the values of P and Z for optimization under $AMIGO_{OBT}$ As follows: P = 0, Z1 = -176.1048, Z2 = - 0.0518305, and G = 4644.6092.

5. Discussion

This section discussed the comparison results between tested tuning methods separately and jointly with proposed methods in terms of time domain specifications as tabulated in **Table 2**, where the evaluation benchmark is based on analyzing

	Method	td(s)	tr(s)	ts(s)	tp(s)	SSt(s)	Pos %	ζ	Iter. No
Classical Tuning	RAT	0.102	0.267	0.935	1.06	1.5	6.24	0.662	
	Z-N	0.094	0.0144	0.181	1.44	0.25	43.8	0.254	
	S IMC	0.0034	0.0234	0.157	1.18	0.35	17.8	0.481	
	CHR	0.0029	0.0168	0.165	1.45	0.3	44.8	0.247	
	AMIGO	0.00491	0.032	0.163	1.23	0.25	22.9	0.4247	
Proposed Technique	RAT_{OBT}	0.000094	0.000382	0.000617	0.999	1.4 ms	Zero	Zero	9
	Z_N_{OBT}	0.00022	0.0015	0.00967	1	0.05	Zero	Zero	2
	$SIMC_{OBT}$	0.0000412	0.000835	0.00555	1.04	0.012	3.87	0.7192	5
	CHR_{OBT}	0.00002536	0.00077	0.0041	1.02	0.015	2.48	0.76202	3
	$AMIGO_{OBT}$	0.000025	0.000378	0.000607	1 msec	1 msec.	Zero	Zero	5

Table 2.
Comparison of time response specifications between tuning methods under study and proposed technique.

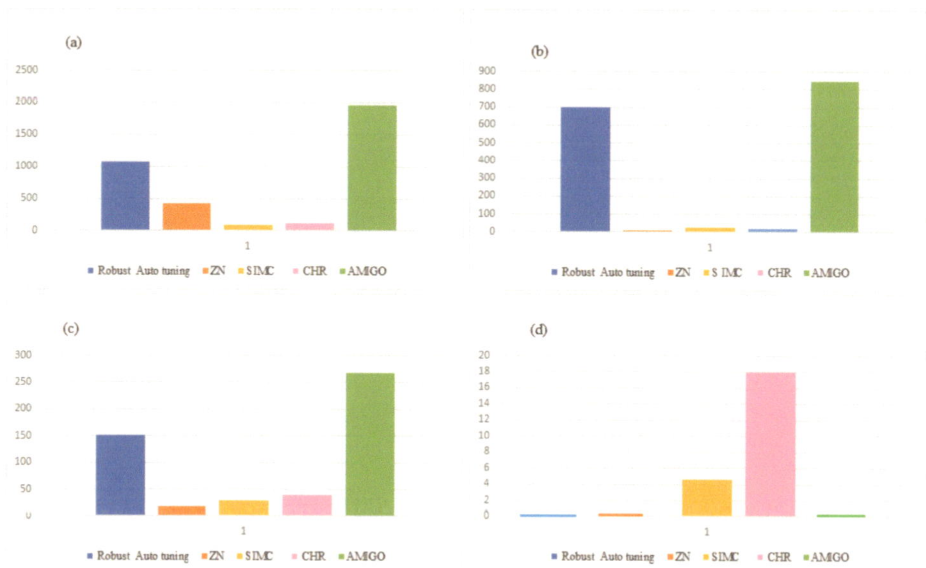

Figure 14.
Comparative improvement time response specifications between tuning methods based OBT *with respect to tuning lonely, (a)improvement td ratio, (b) improvement tr ratio, (c) improvement ts ratio, (d) improvement overshoot ratio.*

Ti and Pos. For the first level tuning, the tuning based Z-N shows that the overshoot is the largest value by 43% between other tuning methods. For the second level optimization, the time response produced by the proposed technique is quite less than by using tuning lonely. Hence, combining both best tuning methods with optimization obtained maximum minimization Ti to the considered system.

Figure 14 presents the comparative improvement time response between tuning methods based proposed technique with respect to tuning lonely, to demonstrates the effectiveness of the proposed technique. The results based $\boldsymbol{RAT_{OBT}}$ show that the improving time response ratio in terms of td, tr, ts duplicated by 1081, 699, 151 times respectively compared with RAT tuning lonely. Where, $\boldsymbol{Z\_N_{OBT}}$ raised by 429, 9.6, 18.7 times respectively. In $\boldsymbol{SIMC_{OBT}}$ boosted by 82.5, 28, 28.2 times respectively, where in $\boldsymbol{CHR_{OBT}}$ increased by 114.3, 21.8, 40.2 times respectively and in $\boldsymbol{AMIGO_{OBT}}$ increased by 1964, 846, 268.5 times respectively. According to the results, it can be observed that $\boldsymbol{AMIGO_{OBT}}$ obtains highest reduction with quit shortest responses td 25 µs, tr 378 µs, ts 607 µs, tp 1 ms, providing better optimization gains Kp 2660, Ki 42400, Kd 13.6. Hence, the time response parameters obtained by the novel proposed technique much smaller and quite acceptable than those using just tuning. It is observed that the eliminating overshoot was achieved in $\boldsymbol{RAT_{OBT}}$, $\boldsymbol{Z\_N_{OBT}}$ and $\boldsymbol{AMIGO_{OBT}}$, where it is quite small in $\boldsymbol{SIMC_{OBT}}$ and $\boldsymbol{CHR_{OBT}}$.

Figure 15 illustrates the comparative time response improvement ratio between $\boldsymbol{AMIGO_{OBT}}$ and other tuning methods based *OBT*. The time response results show that the $\boldsymbol{AMIGO_{OBT}}$ obtains superior performance over $\boldsymbol{Z\_N_{OBT}}$, $\boldsymbol{SIMC_{OBT}}$, $\boldsymbol{CHR_{OBT}}$, and produces a higher reduction time response to be measured in Microsecond unit, which gives the capability to overcome the majority of previous works further. This is typically within the required criteria for robotic applications. The final design with step response based $\boldsymbol{AMIGO_{OBT}}$ was simulated based Matlab/Simulink. **Table 3** refers to parameters of the system in terms of P and Z location and proportional gains in both separately tuning and jointly tuning based proposed method. Accordingly, the compensator TF produced by different methods is compared based on the steady state response and system characteristics. The comparison of modification Z location and adjusting compensator gains was presented to show the effectiveness of using AMIGO tune based proposed methodology over

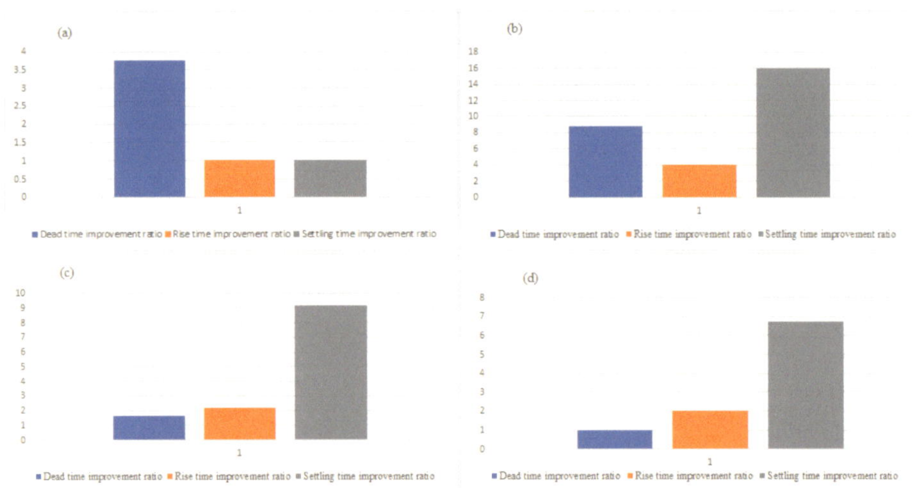

Figure 15.
Comparative time response improvement ratio between $AMIGO_{OBT}$ *and other tuning methods based* OBT, *(a) improvement ratio comparing with* RAT_{OBT}, *(b) improvement ratio comparing with* Z_N_{OBT}, *(c) improvement ts ratio comparing with* $SIMC_{OBT}$, *(d) improvement ratio comparing with* CHR_{OBT} .

	Method	No of Z	No of P	Z location	P location	Kp	Ki	Kd
Classical Tuning	RAT	2 real	1 real	(−69.6), (−6.1)	0	319	1790	4.22
	Z-N	2 real	1	(−96.8), (−96.8)	0	13.92693	10.656875	36.2
	S IMC	2 real	1	(−242.3), (−15.15)	0	5290	75500	20.6
	CHR	2real	1	(−177), (−52.6)	0	553	22400	2.41
	AMIGO	2 real	1	(−177.5), (−17.52)	0	2660	42400	13.6
Proposed Technique Technique	***RAT***$_{OBT}$	2 real	1 real	(−0.07859), (−3.745)	0	4.6888	181	615
	Z_N$_{OBT}$	2 complex	1 real	(−15.85 + 8.95e-07i), (−15.854–8.95e-07i)	0	10.67551	10.6568757	2.5938410
	SIMC$_{OBT}$	2real	1	(−236.9), (−0.1663)	0	454000	75500	1920
	CHR$_{OBT}$	2 real	1	(−177.19), (−52.58)	0	36200	22400	219
	AMIGO$_{OBT}$	2 real	1	(−176.1), (−0.05183)	0	8180	42400	4640

Table 3.
Comparison proportional gains and poles zeros location between tuning methods under study and proposed technique.

another tuning understudy. Accordingly, the compensator TF produced by different methods is compared based on the steady state response and system characteristics.

The damping ratio(ζ) can be specified by solving the Eq. (8). It can be seen that both SIMC and CHR based proposed produced highest damping ratio 0.71920, 0.76202 respectively compared with other cases, where it eliminated in $\boldsymbol{RAT_{OBT}}$ and $\boldsymbol{Z - N_{OBT}}$, $\boldsymbol{AMIGO_{OBT}}$.

$$\zeta = \frac{-ln\left(\frac{Pos}{100}\right)}{\sqrt{\pi^2 + ln^2\left(\frac{Pos}{100}\right)}} \tag{8}$$

By contrast, the proposed method provides better performance and maximum reduction time response to be measured Microsecond unit, providing a magnificent technique to overcome the major previous works in case of the majority of them not accomplish the highest reduction, despite using evolutionary algorithms to solve controller problems as in [42–47].

6. Conclusion

This work presents a very simple analytic tuning procedure, which yields surprisingly superior results and is boosted with improved test bench design. Further, it is well suited to optimize tuning PID parameters. It was proposed a novel technique based on different tuning methods RAT, Z-N, S IMC, CHR, and AMIGO to be combined with optimization SISO application. The performance was analyzed and evaluated in terms of time response specifications, optimal PID gains, and poles zeros location. The time response of all methods and their system characteristics are compared with each other. Further, the designed controller is implemented on a second order system using MATLAB/SISO tool environment. Simulation results demonstrate that all tested tuning based proposed method can enhance the time response considerably, where $\boldsymbol{AMIGO_{OBT}}$ provides greatly improved over existing methods were studied in this paper, particularly for eliminating overshoot and minimization time response. It was proved that using $\boldsymbol{AMIGO_{OBT}}$ can be considered as a novel method to improve the efficiency of the PID controller. The proposed strategy can abolish the drawback in other techniques, next to obtain the best solution to improve the speed performance of the controller. Further providing a capability to improve the robustness, stability, proficiency, besides easily and quickly calculations. The importance of the proposed technique is that the control action provides high precision response and significantly shorten overshoot which is more applicable to be used in a robotic control system for high speed applications. Moreover, it can be used to control a higher order system. Considerably, it gives superb information into how the controller should be returned in response to process changes in terms of proportional gains and time delay. In the scope of the future, it is possible to use this technique with the Genetic Algorithm to optimize the DC motor controller significantly, prepared to configure the hardware on Field Programmable Gate Array System on Chip (FPGA-SoC), using both MATLAB and VIVADO application.

Conflict of interest

The authors declare that there is no conflict interest.

Author details

Falih Salih Mahdi Alkhafaji[1]*, Wan Zuha Wan Hasan[2], Nasri Sulaiman[3] and Maryam Mohd. Isa[3]

1 Head of Laboratory for Electrical and Electronic Engineering Laboratory, Ministry of Industry and Mineral, Iraq

2 Head Center for Biomedical Imaging and Sensor, Faculty of Engineering, Department of Electrical and Electronic Engineering, Universiti Putra Malaysia, Malaysia

3 Department of Electrical and Electronic Engineering, Faculty of Engineering, Universiti Putra Malaysia, Malaysia

*Address all correspondence to: falih_alkafaji1@yahoo.com

References

[1] A. Abdulameer, M. Sulaiman, M. Aras, and D. S. Saleem, "Tuning Methods of PID Controller for DC Motor Speed Control," *Indones. J. Electr. Eng. Comput. Sci.*, vol. 3, no. December, pp. 343–349, 2016.

[2] A. Škraba, V. Stanovov, and E. Semenkin, "Development of control systems kit for study of PID controller in the framework of cyber-physical systems," *IOP Conf. Ser. Mater. Sci. Eng.*, vol. 734, no. 1, pp. 1–8, 2020.

[3] Apekshit B. and Shraddha D., "Comparison of PID Tuning Techniques for Closed Loop Controller of DC-DC Boost Converter," *Int. J. Adv. Eng. Technol.*, vol. 8, no. 1, pp. 2064–2073, 2015.

[4] O. Chao and L. Weixing, "Comparison between PSO and GA for parameters optimization of PID controller," *2006 IEEE Int. Conf. Mechatronics Autom. ICMA 2006*, vol. 2006, pp. 2471–2475, 2006.

[5] K. H. Raut and S. R. Vaishnav, "' Performance Analysis of PID Tuning Techniques based on Time Response specification ,'" *Int. J. Innov. Res. Electr. Electron. Instrum. Control Eng.*, vol. 2, no. 1, pp. 616–619, 2014.

[6] M. Shahrokhi and A. Zomorrodi, "Comparison of PID Controller Tuning Methods," *Proc. 8th Natl. Iran. Chem. Eng. Congr.*, pp. 1–12, 2002.

[7] P. V. G. K. Rao, M. V. Subramanyam, and K. Satyaprasad, "Study on PID controller design and performance based on tuning techniques," *2014 Int. Conf. Control. Instrumentation, Commun. Comput. Technol. ICCICCT 2014*, no. July, pp. 1411–1417, 2014.

[8] O. Aydogdu and M. Korkmaz, "A Simple Approach to Design of Variable Parameter Nonlinear PID Controller," *Int. Conf. Adv. Inf. Technol. with Work. ICBMG*, vol. 20, pp. 81–85, 2011.

[9] A. B. Smitha, S. C. N. Shetty, and B. Baby, "PID Controller Tuning and Its Case Study," *Int. J. Res. Appl. Sci. Eng. Technol.*, vol. 4, no. Vi, pp. 292–299, 2016.

[10] S. Zhao, S. Liu, R. De Keyser, and C. M. Ionescu, "The application of a new PID autotuning method for the steam/water loop in large scale ships," *Processes*, vol. 8, no. 2, 2020.

[11] M. N. Anwar and S. Pan, *Synthesis of the PID controller using desired closed-loop response*. IFAC, 2013.

[12] H. Zhang, Y. Cai, and Y. Chen, "Parameter Optimization of PID Controllers Based on Genetic Algorithm," *2010 Int. Conf. E-Health Networking, Digit. Ecosyst. Technol. Param.*, pp. 47–49, 2010.

[13] V. Vishal, V. Kumar, K. P. S. Rana, and P. Mishra, "Comparative Study of Some Optimization Techniques Applied to DC Motor Control," *2014 IEEE*, pp. 1342–1347, 2014.

[14] I. Chiha, J. Ghabi, Liouane, and N. Ecole, "TUNING PID CONTROLLER WITH MULTI-OBJECTIVE DIFFERENTIAL EVOLUTION," *Proc. 5th Int. Symp. Commun. Control Signal Process. ISCCSP 2012, Rome, Italy, 2–4 May 2012*, no. May, 2012.

[15] F. S. M. Alkhafaji, W. Z. W. Hasan, M. M. Isa, and N. Sulaiman, "Response Time Reduction for DC Motor Controller Using SISO technique," vol. 5, no. 3, pp. 401–408, 2017.

[16] U. T. Tochukwu Livinus, "Effects of PID Controller on a Closed Loop

Feedback System," *Int. J. Sci. Eng. Res.*, vol. 9, no. 4, pp. 1255–1258, 2014.

[17] F. S. M. Alkhafaji, W. Z. W. Hasan, M. M. Isa, and N. Sulaiman, "Robotic Controller : ASIC versus FPGA — A Review," *J. Comput. Theor. Nanosci.*, vol. 15, no. 1, pp. 1–25, 2018.

[18] M. M. R. ALAVI, T. ÇAVDAR, and V. F. AGHJEHKAND, "Particle Swarm Optimization - Based Determination of Ziegler-Nichols Parameters for PID Controller of Brushless DC Motors," 2012.

[19] S. Skogestad, "Simple analytic rules for model reduction and PID controller tuning," *J. Process Control*, vol. 13, no. 4, pp. 291–309, 2003.

[20] S. Skogestad and C. Grimholt, "The SIMC method for smoth PID controller tuning," *Sci. Technol.*, no. Skogestad 2003, pp. 1–29, 2011.

[21] W. Consulting, "Process Control Newsletter," *Spring 2013*, no. 6, 2013.

[22] A. Jayachitra and R. Vinodha, "Genetic Algorithm Based PID Controller Tuning Approach for Continuous Stirred Tank Reactor," *Adv. Artif. Intell.*, vol. 2014, 2014.

[23] I. Fiodorov, "Synthesis Algorithms of Controllers for Automatic Control Systems with Maximum Stability Degree Synthesis Algorithms of Controllers for Automatic Control Systems with Maximum Stability Degree," *Ann. Univ. Craiova, Electr. Eng. Ser.*, no. No. 37, p. 5, 2013.

[24] J. Paulusová and M. Dúbravská, "Application of Design of Pid Controller for Continuous Systems," *Humusoft.Cz*, no. 1, pp. 2–7, 2012.

[25] Z. Haiyang, S. U. N. Yu, L. I. U. Deyuan, and L. I. U. Hao, "Adaptive Neural Network PID Controller Design for Temperature Control in Vacuum Thermal Tests," pp. 458–463, 2016.

[26] P. R. Dasari, K. Raviteja, and A. S. Rao, "Optimal H 2 – IMC based PID Controller Design for Multivariable Unstable Processes," *2017 Indian Control Conf.*, vol. 49, no. 2, pp. 403–408, 2017.

[27] R. S. Naik, P. G. Student, A. Pradesh, and A. Pradesh, "Tuning of PID Controller by Ziegler-Nichols Algorithm for Position Control of DC Motor," *IJISET - Int. J. Innov. Sci. Eng. Technol.*, vol. 1, no. 3, pp. 379–382, 2014.

[28] E. E. Vladu and T. L. Dragomir, "Controller Tuning Using Genetic Algorithms," *Proc 1st Rom. Jt. Symp. Appl. Comput. Intell.*, pp. 1–10, 2004.

[29] A. Nayak and M. Singh, "Study of Tuning of PID Controller By Using Particle Swarm Optimization," *Int. J. Adv. Eng. Res. Stud.*, vol. IV, no. Jan.-March, pp. 346–350, 2015.

[30] N. Kuyvenhoven, "PID Tuning Methods An Automatic PID Tuning Study with MathCad," *Neil Kuyvenhoven Calvin Coll. ENGR. 315*, pp. 1–8, 2002.

[31] S. M. and G. S. I.Haasanzadah, "Design and Implementation of a Controller Magnetic Levitation System Using Genetic Algorithms," *J. Appl. scinces*, vol. 8, no. 24, pp. 4644–4649, 2008.

[32] G. Mantri and N. R. Kulkarni, "Design and Optimization of Pid Controller Using Genetic Algorithm," *Int. J. Res. Eng. Technol.*, vol. 2, no. 6, pp. 926–930, 2013.

[33] L. Ma and M. Wang, "The Optimizing Design of Wheeled Robot Tracking System by PID Control Algorithm Based On BP Neural Network," *Int. Conf. Ind. Informatics - Comput. Technol. Intell. Technol. Ind. Inf. Integr.*, pp. 34–39, 2016.

[34] K. W. G. Michael, "Auto-Tuning: From Ziegler-Nichols To Model Based Rules," *J. Chem. Inf. Model.*, vol. 53, p. 160, 1989.

[35] L. X. F. Wu R., Zhang W., "The IMC-PID Controller Design for TITO Process Using Closed-loop Identification Method," *2014 13th Int. Conf. Control. Autom. Robot. Vis. Mar. Bay Sands, Singapore*, no. December, pp. 1339–1344, 2014.

[36] K. Raviteja, P. R. Dasari, and A. S. Rao, "Improved controller design for two-input-two-output (TITO) unstable processes," *Resour. Technol.*, vol. 2, pp. S76–S86, 2016.

[37] MathWorks, "Robust Tuning Approaches," 2020. [Online]. Available: https://www.mathworks.com/help/robust/gs/robust-tuning-workflows.html.

[38] T. H. Astrom K J, "Revisiting the Ziegler – Nichols step response method for PID control," *J. Process Control*, vol. 14, pp. 635–650, 2004.

[39] M. Shamsuzzoha and M. Lee, "IMC Filter Design for PID Controller Tuning of Time Delayed Processes," pp. 253–286, 2012.

[40] Staya Sheel and Omhari Gupta, "New Techniques of PID Controller Tuning of a DC Motor—Development of a Toolbox," *MIT Int. J. Electr. Instrum. Eng.*, vol. 2, no. 2, pp. 65–69, 2012.

[41] K. Astrom, "PID controllers: Theory, Design and Tuning," *Instrument Society of America*. p. 343, 1995.

[42] M. A. Shamseldin, M. A. Eissa, and A. A. El-samahy, "Practical Implementation of GA-Based PID Controller for Brushless DC Motor," *17th Int. Middle East Power Syst. Conf. Mansoura Univ. December 15–17 ,2015*, no. December, 2015.

[43] T. Vijayakumar, S. Muthukrishnan, and G. Murugananth, "Genetic Algorithm Based Speed Control of PMDC Motor Using Low Cost PIC 16F877A Microcontroller," *Circuits Syst.*, vol. 07, no. 08, pp. 1334–1340, 2016.

[44] B. Bourouba and S. Ladaci, "Comparative performance analysis of GA, PSO, CA and ABC algorithms for fractional PIλDμ controller tuning," *Proc. 2016 8th Int. Conf. Model. Identif. Control. ICMIC 2016*, no. 2, pp. 960–965, 2017.

[45] S. K. Suman and V. K. Giri, "Speed Control of DC Motor Using Optimization Techniques Based PID Controller," in *2nd IEEE International Conference on Engineering and Technology (ICETECH)*, 2016.

[46] W. M. Elsrogy, N. K. Bahgaat, M. El Sayed, and M. M. Hassan, "Speed Control of DC Motor Using PID Controller Based on Changed Intelligence Techniques," *Int. J. Swarm Intell. Evol. Comput.*, vol. 07, no. 01, 2018.

[47] S. K. Raja, V. P. Badathala, and S. S. K, "LFC Problem by Using Improved Genetic Algorithm Tuning PID Controller," *Int. J. Pure Appl. Math.*, vol. 120, no. 6, pp. 7899–7908, 2018.

Chapter 3

Highway PC Bridge Inspection by 3.95 MeV X-Ray/Neutron Source

Mitsuru Uesaka, Katsuhiro Dobashi, Yuki Mitsuya, Jian Yang and Joichi Kusano

Abstract

We have developed portable 950 keV/3.95 MeV X-ray/neutron sources and applied them to inspection of PC concrete thicker than 200 mm within reasonable measuring time of seconds - minutes. T-girder-, Box- and slab- bridges are considered. Now we are to start X-ray transmission inspection for highway PC bridge (box) by using 3.95 MeV X-ray sources in Japan in 2020. By obtaining X-ray transmission images of no-grout-filling in PC sheath and thinning of PC wires, we plan to carry out numerical structural analysis to evaluate the degradation of strength. Finally, we are going to propose a technical guideline of nondestructive evaluation (NDE) of PC bridges by taking account of both X-ray inspection and structural analysis. Further, we are trying to detect rainwater detection in PC sheath, and asphalt and floor slab by the 3.95 MeV neutron source. This is expected to be an early degradation inspection. We have done preliminary experiments on X-ray transmission imaging of PC wires and on-grout-filling in the same height PCs in 450–750 mm thick concretes. Moreover, neutron back scattering detection of water in PC sheath is also explained.

Keywords: on-site bridge inspection, highway PC bridge, 950 keV/3.95 MeV X-ray sources, 3.95 MeV neutron source, non-grout-filling, PC wires thinning, rain-water-detection, structural analysis, guideline of nondestructive evaluation

1. Introduction

Since PC (Pre-stressed) bridges were first constructed, about 40 years has passed. Then, the degradation and problems such as no-grout-filling, rail-water-invasion, corruption, thinning and disconnection of PC bridges has revealed and been detected. In these tens years, several nondestructive and destructive evaluations have been performed. Depending on the seriousness of degradation, a few PC bridges are under reconstruction and its planning. So far, they have mainly used inspection by eyes and hammering. However, it is very hard to evaluate the problems and degradation of PC by the above methods. There are non-destructive methods to check PC such as ultrasonography, radar, X-ray transmission by X-ray tubes (200–400 kV), magnetic field detection, etc. But, practically it is difficult to evaluate PC in thicker concrete than 200 mm. We have developed portable 950 keV/3.95 MeV electron linac (linear accelerator)-based X-ray / neutron sources and applied them to inspection of PC concrete thicker than 200 mm within reasonable measuring time of seconds – minutes more than 10 times so far [1–4]. Major poor-construction and degradation of PC bridges to be detected by X-rays and neutrons are unfilled grout, rainwater intrusion, corrosion and thinning and disconnection of PC wires. Unfilled grout and thinning and disconnection of PC wires can be measured by difference of X-ray attenuation coefficient by X-rays. The three types of the PC bridges and locations of X-ray/neutron source and detector are depicted in **Figure 1**. As for highway bridges, concrete vertical WEB wall is thick as 450–1,000 mm.

There are of course many nondestructive evaluation (NDE) methods to try to detect poor construction such unfilled grout in PC sheath and degradation such as thinning and disconnection of PC wires. RADAR, ultrasonic testing, magnetic testing, 200–400 kV X-ray tube, etc. are candidates as shown in **Figure 2**. However, they are available for thinner concrete than 200 mm. On the other hand, 950 keV / 3.95 MeV X-ray sources can be used for transmission testing for thick concrete of 200–400 mm and 200–1,000 mm based on the calculation and our experience so far [1–4]. We think that only the 950 keV/3.95 MeV X-ray sources can enable transmission imaging of PC structures in 200–1,000 mm thick concrete within minutes.

Figure 3 summarizes major poor construction of unfilled grout, early degradation such as rainwater intrusion and finally serious degradation of thinning and disconnection, and suitable NDE methods and successive structural analysis. If there is unfilled grout in PC sheath, rainwater may intrude from the edges and is stored there. Then, the rainwater makes PC wires and sheath wall be corroded. Volume expansion of PC sheath due to corrosion and oxidation induces cracks in near concrete. The cracks gradually become larger and reach the concrete surface, When the sheath wall is broken, the rainwater exude out and intrude into concrete and surface. As the corrosion is enhanced, superficial oxidized iron leaves and thinning of PC wires occurs. Even, the disconnection may happen since the wires are tensile. Among the above temporal change of state, the structural changes such as unfilled grout and thinning/disconnection of wires can been detected by X-ray transmission imaging. On the other hand, material property change such as rainwater intrusion can be detected by neutron scattering in water. Since concrete vertical WEB wall is thicker than ~450 mm, 3.95 MeV X-ray source is appropriate with respect to transmission ability. Iron components of PC such as wires can be clearly seen with good contrast to concrete. Their thinning and disconnection are observed with the spatial resolution of 1 mm. Since tension is added to PC wires at its construction, they tend to be attached to the upper wall of PC sheath. Therefore, unfilled grout should be recognized under PC wires with certain change of contrast. Measured flaws such as unfilled grout and thinning/disconnection are inputted to structural analysis described in Chapter 5. Thus, initial poor construction of unfilled grout and serious thinning and disconnection of PC wires are diagnosed and then their affect to

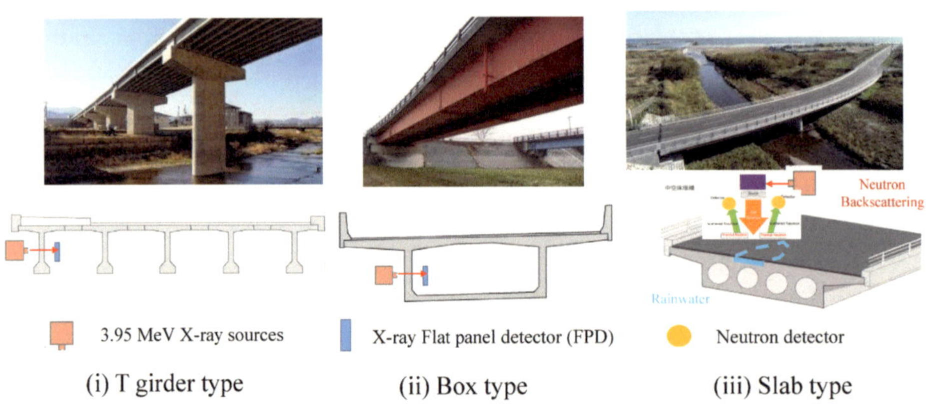

Figure 1.
On-site X-ray/neutron inspection by 3.95 MeV system for three types of highway PC bridge.

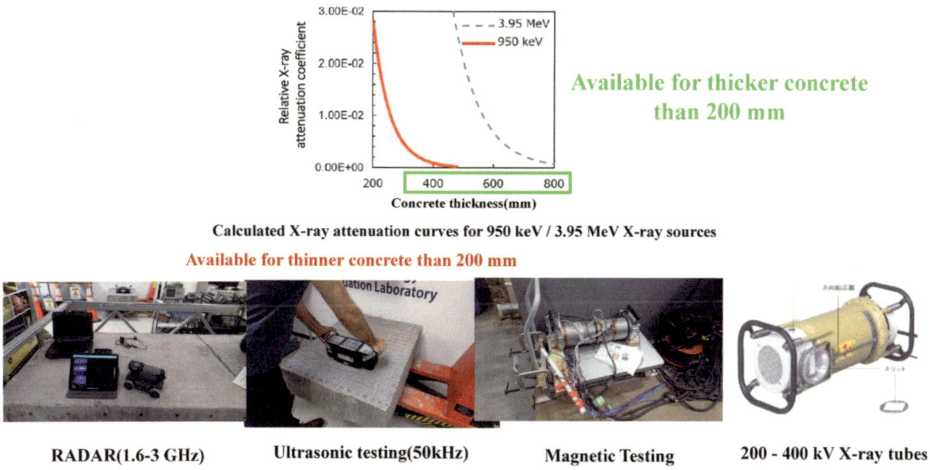

Figure 2.
Advantage of 950 keV/3.95 MeV X-ray sources over other NDE methods for thicker concrete than 200 mm.

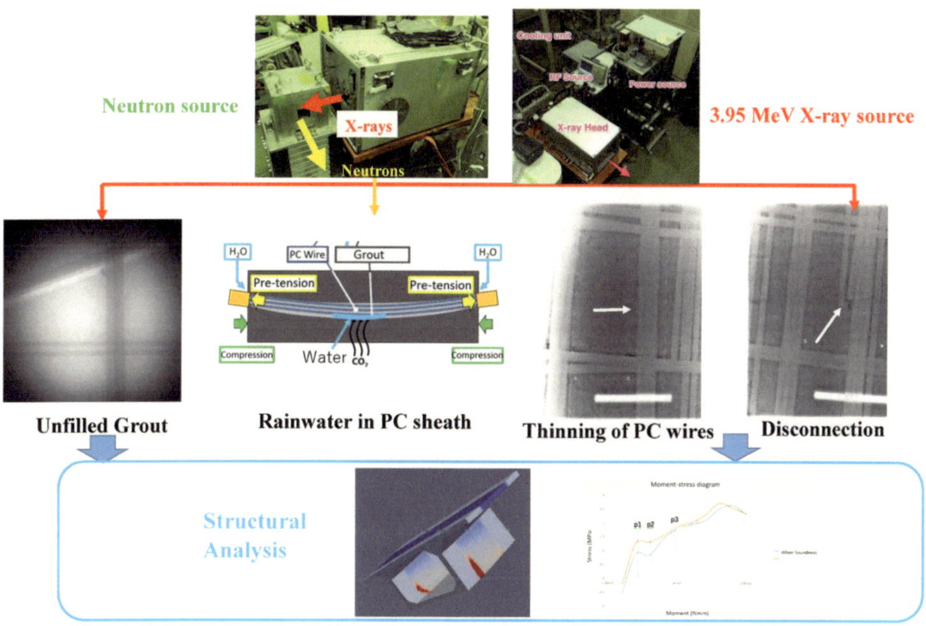

Figure 3.
Poor construction and degradation of PC bridges, suitable NDE methods and structural analysis.

lack and degradation of strength is quantitatively evaluated by structural analysis. Finally, maintenance, repairing and reconstruction are planned. Moreover, there is a possibility to detect rainwater intrusion inside PC sheath by portable 3.95 MeV neutron source [5], which is discussed in Section 3.2.

2. 3.95 MeV electron X-ray/neutron sources

We use X-band (9.3 GHz) linac based 3.95 MeV X-ray sources for the inspection of the actual bridge [1–3]. The systems are shown in **Figure 4**. The electrons are accelerated up to 3.95 MeV by radio frequency (RF) fields. We also adopted the side-coupled standing wave type accelerating structure. Electrons are injected into

Specification	
RF frequency	9.3 GHz
Beam focusing	RF focusing
X-ray Intensity	2 Gy/min at 1 m
RF source	9.3 GHz Magnetron 1.3MW, Pulse width 4μs Repetition rate 200 pps
Electron gun	Voltage 35kV Pulse Peak Current 300 mA
X-ray head weight	200 kg (with Pd shielding)
RF unit weight	100 kg
Power source	120 kg
Water chiller unit	100 kg

Figure 4.
3 95 MeV portable X-band linac based X-ray source and its major parameters. The system consists of four units: X-ray head, magnetron, power, and chiller units.

a Tungsten target that generates bremsstrahlung X-rays. The generated X-rays are collimated by a Tungsten collimator into the shape of a cone which has an opening angle of 17 degrees. Most important is the X-ray intensity, which is 2 Gy/min at 1 m for a full magnetron RF power of 1.8 MW. The system consists of a 200 kg (including local Pb radiation shielding) X-ray head, 100 kg magnetron box, and stationary electric power source and water chiller unit. The X-ray head and magnetron box are portable, and because they are connected to each other by a flexible waveguide, only the position and angle of the X-ray head are finely tuned. We have optimized the design with respect to X-ray intensity, compactness, and weight. The parameters of the 3.95 MeV X-ray source are summarized in the table in the figure.

We place an X-ray detector on the opposite site of the X-ray source between the object and source to detect the transmitted X-rays through the object. We use a flat panel detector (FPD) by Varian Co. for the detector. The detect and its specification are given in **Figure 5.**

In order to be able to perform neutron TOF measurement, a pulsed neutron source is needed. Usually this kind of neutron source were produced by large-sized, high-energy particle linear accelerator, but by using X-Band type electron linac which compensate its small size with high frequency, the size of the neutron source can be reduced greatly and even possible for mobility.

	XRD 1622 AO19 IMG
Detector Size	16" × 16" (41cm × 41cm)
Energy Range	20keV-15MeV
Scintillator Type	Gd_2O_2S:Tb & Cu Filter PI-200 (Mitsubishi Chemical)
Resolution	200μm Pixel Size
Phosphor Layer	436μm, 200mg/cm^2
Frame Rate	1 fps
Electronics	14bit ADC & 2Gain Settings
Interface	Gigabit Ether Net (GigE)
Weight	8.8kg

Figure 5.
X-ray flat panel detector (FPD) and its specification.

When we put a Beryllium target in from of the 3.95 MeV X-ray source, it becomes a neutron source, too (see **Figure 6**). ^{9}Be, having the lowest threshold energy for photo-nuclear reaction $^{9}Be(\gamma, n)^{8}Be$ to generate neutrons. A beryllium photo-neutron target (50 mm × 50 mm × 50 mm) has been combined with a lead beam collimator, a boric acid resin layer for neutron shielding, and a lead layer for X-ray shielding. Since mainly fast neutrons are used in the neutron source, a beam line using a high Z material that does not moderate the neutrons is used. Optimization of the beryllium target size and neutron/X -ray shielding simulation is performed using the Monte-Carlo code. The target weight is about 100 kg.

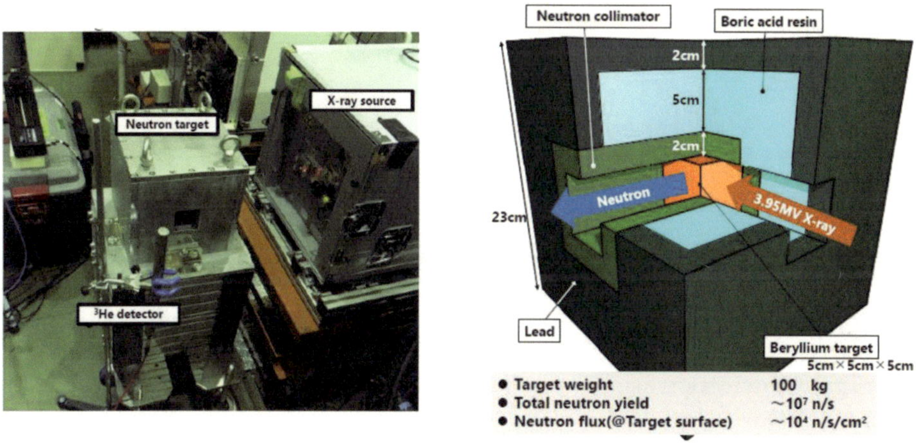

(i) (ii)

Figure 6.
Neutron source by neutron target in front of 3.95 MeV X-ray source. (i) Photograph (ii) inner structure.

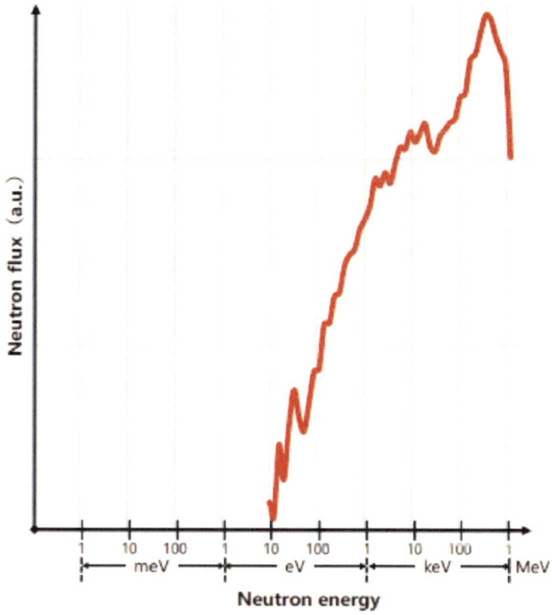

Figure 7.
Produced neutron energy distribution measured by 3He gas detector and TOF method and energy range of neutrons scattered by water.

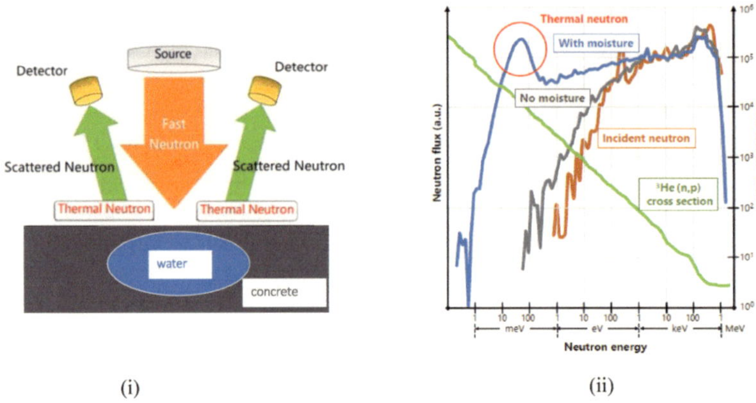

Figure 8.
Neutron energy profiles of ^{3}He gas (n,p) cross section, incident neutrons and backscattered neutrons from concrete with and without water cell. (i) Neutrons scattered by water in concrete. (ii) energies of incident and scattered neutrons in water and ^{3}He gas detector efficiency.

The calculated neutron yield in the neutron source is approximately ~10^7 n/s (neutrons/second), which is more intense than ~10^6 n/s of ^{252}Cf (1 μg) moister detector used for NDE in chemical plants. The distribution of neutron energy produced is measured by ^{3}He gas detector and TOF (Time of Flight) method, as shown in **Figure 7**.

Rainwater detection using the 3.95 MeV neutron source is performed by irradiating concrete with fast neutrons and detecting backscattered moderated neutrons due to multiple elastic scattering with light elements especially hydrogen nuclei (see **Figure 8(i)**). Neutron detection using the ^{3}He gas is attributed to high reaction cross section with neutrons in the thermal region. Therefore, the count of detectors increases with the existence of water as shown in the figure.

3. X-ray transmission detection and evaluation for highway bridge by 3.95 MeV X-ray source

We plan to do NDE by 3.95 MeV X-ray source for highway PC bridge in 2020. Its goal is to detect and visualize unfilled grout via X-ray transmission imaging. **Figure 9** shows transvers and longitudinal cross sections, possible location of unfilled grout and typical X-ray transmission images by 950 kV source for filled and unfilled grouts. Since tensile PC wires tends to attach the upper inner surface in PC sheath, the grout is filled in the lower space there. Therefore, in general, unfilled grout occurs in the lower space (see (i), (ii)). In construction stage, grout is pushed and filled from one side of sheaths so that unfilled grout tends to occur at an ascending part of the opposite side as shown in (iii). The unfilled grout in PC sheathe in 200–250 mm thick T girder WEB was measured and visualized at the ascending part as show in (iv). The unfilled part looks white comparing gray or black part of filling. Phase of unfilled can be evaluated by quantitative evaluation of gray value, namely X-ray attenuation coefficient. Appearing X-ray transmission imaging of filled and unfilled grout would be a goal of the coming task for highway PC bridges.

Here we consider PC bridges of box type and T girder Typical cross section, location of PC sheath, X-ray head and detector are summarized in **Figure 10**. X-ray head is movable 200 kg weigh component among the four. It can be accessed to the WEB with RF source via flexible waveguide for RF power delivery for electron beam acceleration. For example, the thickness of WEB of Box type and T girder are

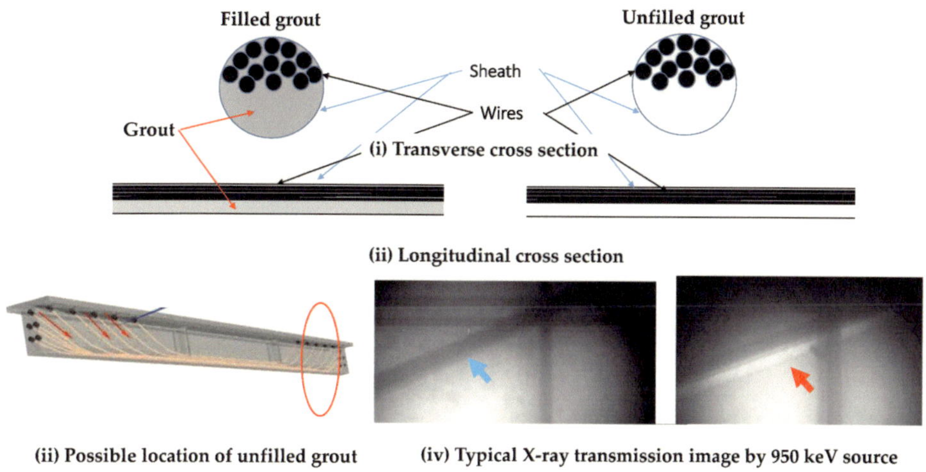

Figure 9.
Transvers and longitudinal cross sections, possible and typical X-ray transmission images by 950 kV source for filled and unfilled grouts.

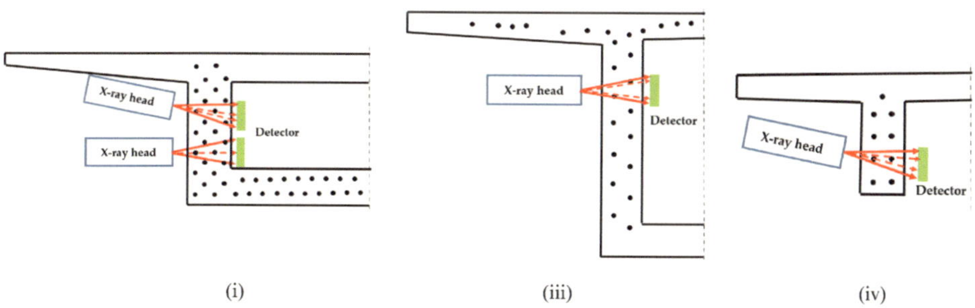

Figure 10.
Typical WEBs and locations of X-ray source and detector for box-type and T. (i) Box thick WEB. (ii) box thin WEB. (iii) T-girder WEB.

750, 550 and 450 mm, respectively. Closed circles represent typical locations of PC sheaths. Vertical arrays of PC sheaths are arranged in the zigzag way ((i), (ii)) and at the same vertical level ((iii)). When we allocate the X-ray head and eject X-rays horizontally, PC sheaths can form separate images at the detector for two zigzag arrays (see (ii)). However, in the lower case of (i) two images of the PC sheaths at the same vertical level are overlapped at the detector. In order to obtain two transmitted images for the above case, we allocate the X-ray head higher by about 100 mm and decline it by about 10 degrees as shown in the upper case of (i) and (iii). WE have almost finalized the way of access and allocation of the X-ray head to all cases as shown in the figure.

We have carried out the simulating experiments for real highway bridges inspection in the configuration of **Figure 10(i)–(iii)**. In order to simulate the concrete thickness and number and location of PC sheaths, we use cut samples from real old PC bridges which were deconstructed after finishing their roles as shown in **Figure 11(i)**.

First, we explain the results of separate X-ray transmission images for two PC sheaths at different vertical level as shown in **Figure 11(ii)**. Total concrete thickness is 750 mm and 10 iron PC wires of $70m^{\phi}$ are inserted to vacant Sheath 1 and 2. X-ray shot duration is 10 s and it is summed by 100 times. Original image is (ii) and (iii) is the boundary enhanced image by the principle of local contrast enhancement using gray level of the neighborhood pixels [6]. The two PC sheaths and wires are clearly

Figure 11.
Separate X-ray transmission images for two PC sheaths at different vertical level.

recognized separately in both images. In the boundary enhanced image, the sheath and unfilled grout looks clearer than the original, although its background becomes noisy. Instead, the original image is totally ambiguous, but the sheaths wires can be recognized.

Second, we measured three PC sheaths at the same vertical level and horizontal X-ray ejection as shown in **Figure 12**. From the left X-ray source as (i), (ii), 10 wires are inserted in the lower space of the first PC sheath 1. Grout remains in the half space of the second PC sheath 2. The third PC sheath is vacant. X-ray transmission images were obtained by 10 s duration times 1 and 100 as shown in (iii), (iv), respectively. There is almost no remarkable difference as to the quality of image. Only wires in the lower space and upper vacant space, which corresponds to unfilled grout, in the PC sheath 1 can be seen and the PC sheaths 2, 3 are hidden behind it. Since PC wires are attached to the upper inner surface of sheath in a real case, real image is upside down of these images.

Third, the configuration and transmission images for three same vertical level PC sheaths and 100 mm up −10 degree declining X-ray ejection are given in **Figure 13**. Here we try to get separate images for the three same vertical level PC sheaths at the detector. Images by 10 s duration times 1 and 100 shots are shown in (ii) and (iii), respectively. Again there is no remarkable difference with respective to image quality between them. Here the three PC sheaths are becoming separate, but still partially overlapped. We can surely recognize that PC wires are inserted from the right hand side and stop in front of the left edge in the nearest PC sheath 1 to the X-ray source. Over the PC sheath 1, we can see the PC sheath 2 with partially filled grout. The oblique edge of grout is seen in the left hand side. Finally, vacant PC sheath 3 is observed over the PC sheath 2.

Figure 12.
Image of three PC sheath at the same vertical level and X-ray horizontal ejection.

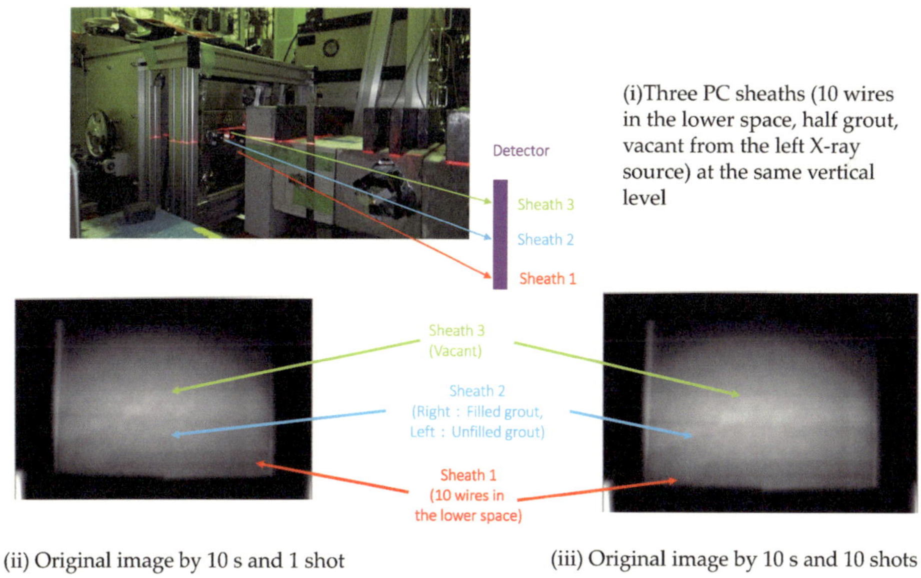

Figure 13.
Configuration and transmission images of three PC sheaths at the same vertical level and 100 mm level up and – 10 degree declining ejection of X-ray source.

Now the findings are summarized in the following.

- Transmission image of 750 mm thick concrete is completely obtained in 10 s by 3.95 MeV X-ray source. We should understand this fact comparing to the calculated X-ray attenuation in the upper graph of **Figure 2**.

- In any cases, PC iron wires can be easily found due to their rather black images. Then, the existence and location of PC sheath is detected. Furthermore, thinning and disconnection can be analyzed.

- In case of several same vertical level PC sheaths and horizontal X-ray ejection, only nearest PC sheath is clearly seen and others are hidden behind it.

- In case of several same vertical level PC sheaths and upward shifted and declining X-ray ejection, they form separate images. Thus, we can detect and evaluate the far side PC sheath.

- Unfilled grout can be evaluated by gray value with respect to rather black image of PC wires in positive image processing. We may be able to form a correlation between measured relative gray value and stage of unfilled grout.

4. Rainwater detection in PC sheath by 3.95 MeV neutron source

Now we are going to explain detection of rainwater intrusion by neutrons. Images of rainwater intrusion in PC bridge and slab type bridge are depicted in **Figure 14**. Rainwater intrusion at unfilled grout in PC sheath causes corrosion of wires to thinning and disconnection, and finally degradation of strength of bridge. We try to detect them by the 3.95 MeV neutron source and ^{3}He gas detector via neutron backscattering. It is expected to be monitoring of early degradation of bridge strength before corrosion of PC wires.

Here we have just started basic experiment on water detection by neutron backscattering considering the situations depicted in **Figure 14**. **Figure 15** shows the experimental configuration. We just put a 50 mm thick bottle of water on a T girder concrete sample cut from a real old bridge. Neutrons from the 3.95 MeV source are scattered in water in the bottle. Backscattered neutrons are detected by the ^{3}He detector, and the TOF method is applied to measure the time of flight from the scattering point to the detector. Flight distance divided by the measured time becomes the velocity of neutron, v and its kinetic energy is obtained as, $\frac{1}{2}mv^2$, for nonrelativistic case.

Measured results of neutron counts as a function of TOF and converted energy are given in **Figures 16(i)** and **(ii)**, respectively. Neutron counts as a function of measurement time and converted neutron energy in cases of unfilled and filled water are plotted in (i) and (ii), respectively. Since the neutron source with about 10^7 n/s is not necessarily intense, the difference with/without water is not remarkable. But we can clearly observe component of neutron backscattering below 1 eV as predicted in **Figure 8**. It can be understood that Proof-of-Principle has been verified.

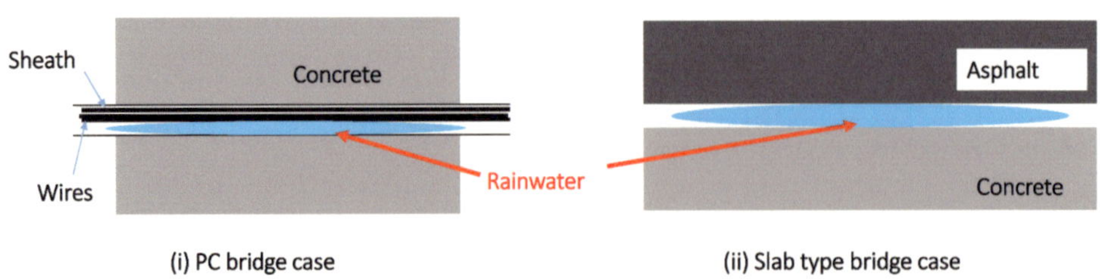

Figure 14.
Images of rainwater intrusion in PC bridge and slab type bridge.

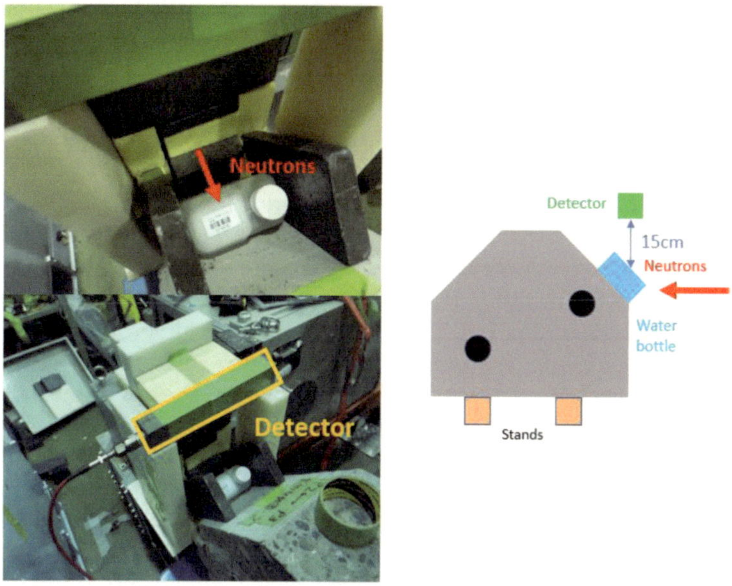

Figure 15.
Experimental configuration modeling rainwater detection by neutron scattering.

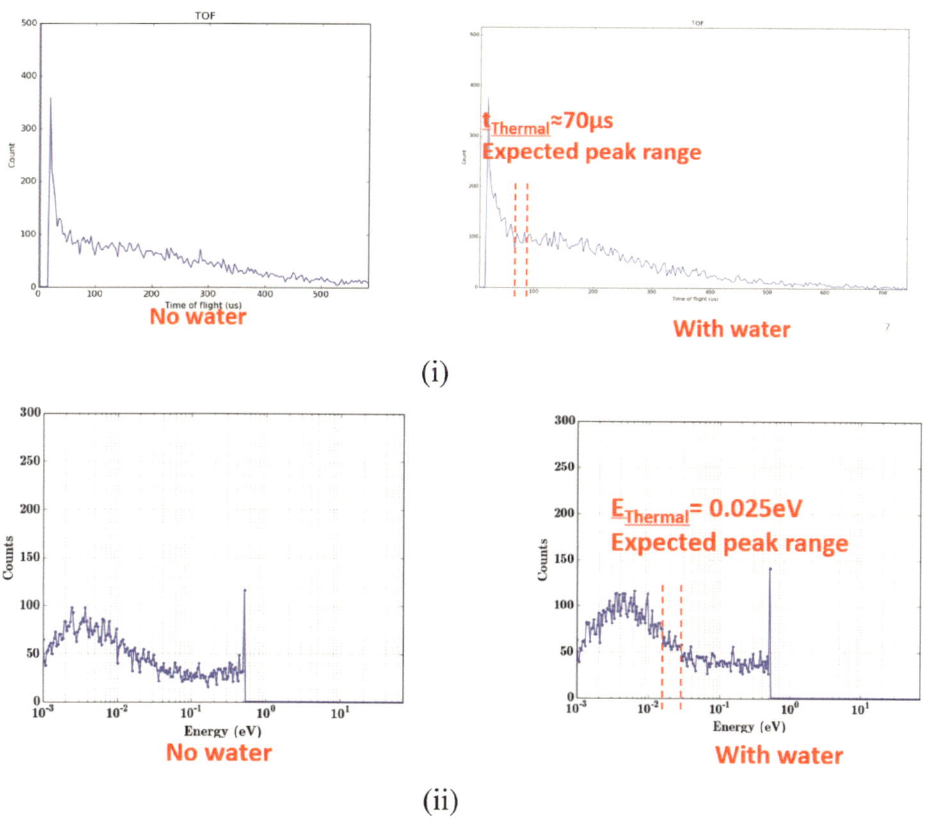

Figure 16.
Comparison of measured backscattered neutron count as a function of TOF with and without water. (i) Neutron counts as a function of TOF. (ii) neutron energy spectra.

As a next step, we are to perform numerical analysis using Monte Carlo code considering the case to detect rainwater in the configuration for slab bridge type. Multilayer sample with asphalt (~75 mm thick), water (more than few mm thick) and concrete (very thick) is first adopted.

5. 3D structural analysis using finite element method

In order to accurately reflect the influence of more detailed degradation conditions obtained by X-ray imaging on structural performance, three-dimensional finite element analysis is required. This chapter aims to evaluate structural strength degradation more precisely by performing structural calculation using finite element analysis software, Du COM-COM3 [7], that can simulate the nonlinear behavior peculiar to concrete with high accuracy. Effect of degradation is evaluated through stress distribution in the cross section of the bridge obtained by the analysis.

The software used for the analysis is DuCOM-COM3 which continues to be developed by the Concrete Laboratory, Department of Civil Engineering, Faculty of Engineering, University of Tokyo [7]. DuCOM-COM3 is used to calculate the mechanical behavior of structures with multi-spans from dynamic characteristics such as earthquake motion and wind vibration to long-term deformation behavior over several decades. DuCOM, which is responsible for the calculation of movement and material deterioration, is coupled to construct structures at various scales from the molecular-scale microscale to the structure-level macroscale.

We introduce an example to use DuCOM-COM3 for evaluation of strength degradation due to measured PC flaws. **Figure 17** shows X-ray transmission images of thinning and disconnection of PC wires of a box type bridge and evaluation of cross section reduction. Degradation of this bridge due to salty water used in winter is rather serious. At the most serious cross section, several PC sheaths are broken and PC wires are heavily thinned and disconnected. We try to evaluate reduction of PC wire cross section as shown in the table of the figure. Then, we input the data to DuCOM-COM3 analysis. **Figure 17** shows one example of box type bridge.

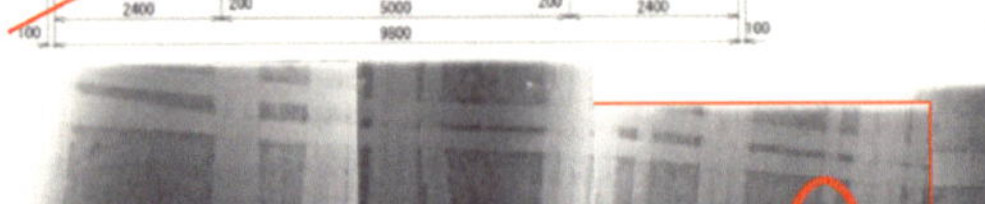

PC Number	Cross section reduction
No.28	100%
No.27	7%
No.26	10%
No.24	10%
No.23	10%
No.21	50%

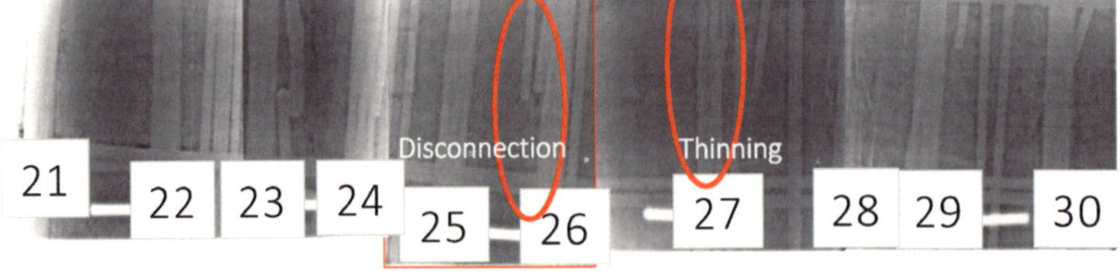

Figure 17.
X-ray transmission images of thinning and disconnection of PC wires of a box type bridge and evaluation of cross section reduction.

We consider a partial block of one span of a four-span PC bridge (see **Figure 17**). We measured most degraded floor cross section by using our X-ray source. Several X-ray transmission images were obtained as shown in the figure. Due to long term corrosion due to salty water, several PC sheaths are broken and some PC wires are thinned and disconnected. Reduction of PC wires cross section is approximately evaluated as shown in the figure. Those data are used for the structural analysis.

3D mesh model is depicted in **Figure 18**. The whole mesh structure corresponds to a part of one span of a four span box type PC bridge. We adopt specialized boundary condition and standard vertical load. Then, we calculate 3D distribution of bending moment, stress, strain and displacement.

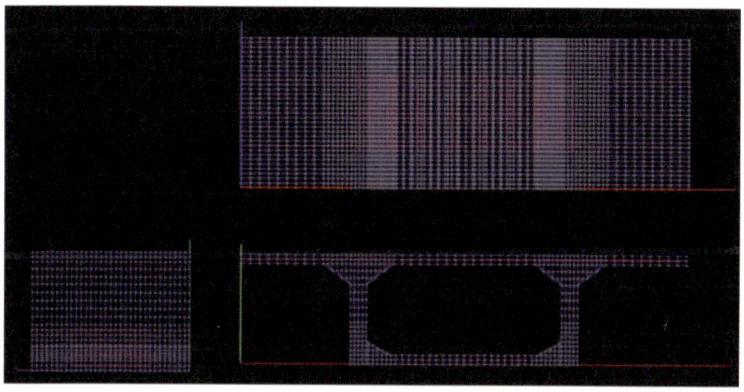

Figure 18.
3D mesh model for the box type bridge with cross section reduction of PC wires at a certain cross section of the bridge.

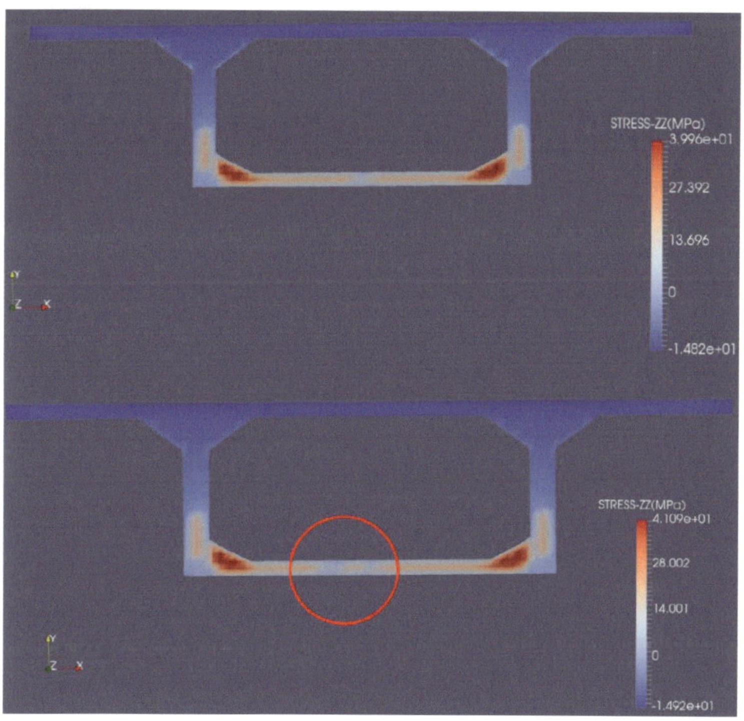

Figure 19.
Comparison of 3D stress distribution of initial (healthy) and degraded states. Circle indicates the degraded part.

3D stress distribution of initial (healthy) and degraded states are given in the upper and lower images of **Figure 19**, respectively We can observe discontinuous distribution around the degraded part indicated by the circle. **Figure 20** also shows the fracture points confirmed by the overall 3D model in this model and analysis.

Figures 21 and **22** show the moment-stress distribution and moment-strain distribution at the lower edge of the bridge section for the initial and degraded states. In both cases, the horizontal axis is the moment, and the vertical axis is the stress and strain. If we can eliminate the moment, we can get standard stress – strain relation. Anyway, we can clearly observe yield phenomenon and yield stress. The upper and lower curves represent the results for the initial and degraded states. Reduction of yield stress is clearly seen due to the degradation. The reduction is 5% in this case. This reduction would be rather serious for the maintenance of the bridge. Actually, it has been decided that this bridge should be reconstructed.

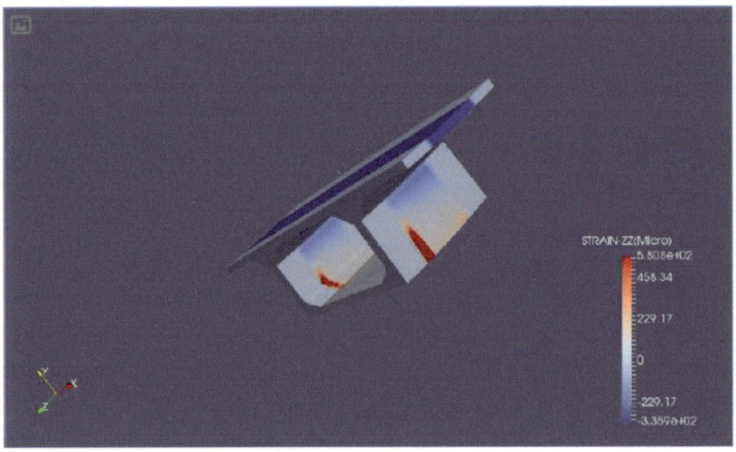

Figure 20.
Confirmation of fracture site in the 3D model.

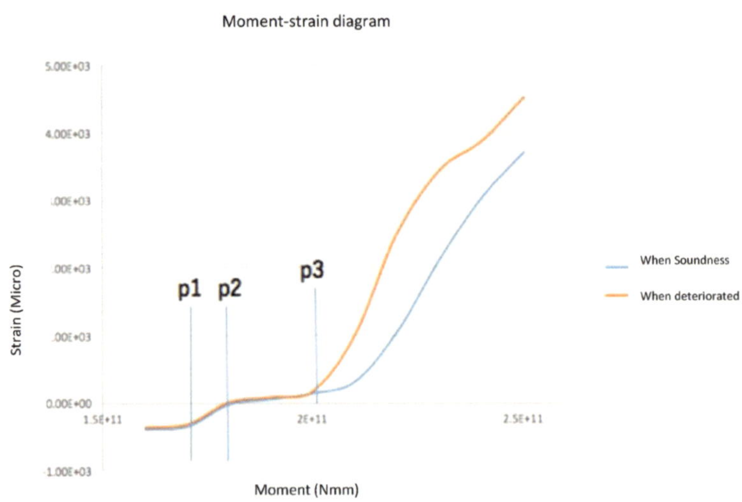

Figure 21.
Moment-stress graph at bottom edge of bridge section.

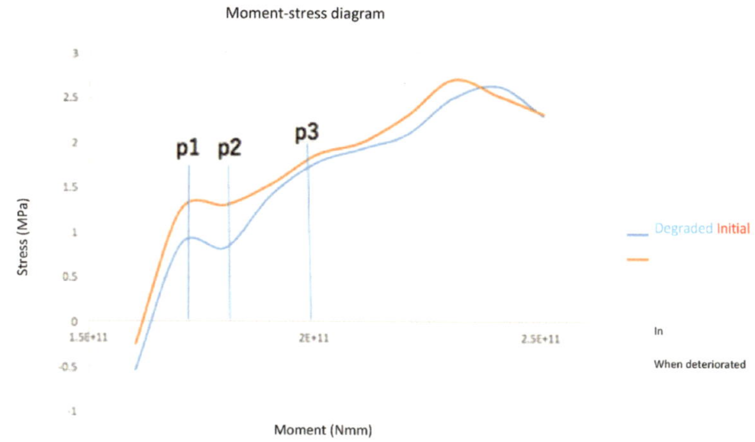

Figure 22.
Moment-strain graph at bottom edge of bridge section.

6. Guidelines for special inspections using 950 keV/3.95 MeV X-ray sources

The Public Works Research Institute and the University of Tokyo are developing new technical guidelines for special inspections of bridges using 950 keV/3.95 MeV X-ray sources. An overview is provided in **Figure 23** [4]. First, visual and hammer sound inspection screening should be performed based on regular inspection guidelines. Advanced hardware and software techniques such as drawn and acoustic analysis are adopted in this step. If degraded parts are found, the special X-ray transmission inspection is performed using the 950 keV or 3.95 MeV X-ray sources, depending on the thickness of the concrete containing the degraded parts (see **Figure 2**). Here, the states of PC wires such as unfilled grout and thinning/disconnection are quantitatively evaluated with spatial resolution of 1 mm. Especially, the state of unfilled grout can be evaluated quantitatively by measuring

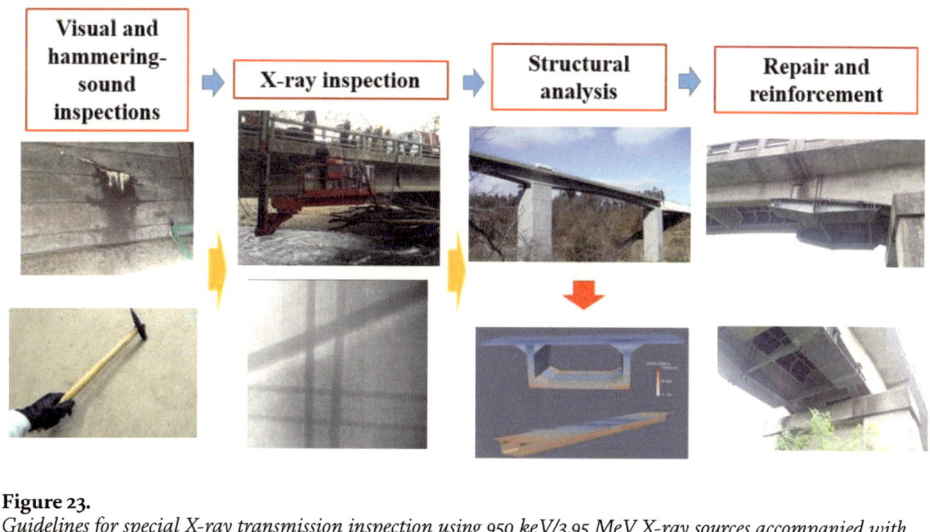

Figure 23.
Guidelines for special X-ray transmission inspection using 950 keV/3.95 MeV X-ray sources accompanied with visual and hammering-sound inspections, structural analysis, final repair, and/or reinforcement.

the gray value, namely the X-ray attenuation coefficient, under the criterion from 0 (white: vacant) through 1 (black as PC wire) (see **Figure 24**). NDE of unfilled grout by γ-emitter radioisotope using such a phase control and decision to proceed to destructive evaluation to obtain more precise information are under way by IFSTTAR (French Institute of Science and Technology for Transport Development and Networks) in France [8]. Then, we plan to perform 3D structural analysis to evaluate the degradation of the structural strength quantitatively as shown in **Figure 25**. We have introduced one example of the structural analysis and evaluation of reduction of yield stress due to thinning and disconnection of PC wires by using DuCOM-COM3 in Chapter 5. We propose regular X-ray transmission inspection and structural analysis every five years and record the change of yield stress. In order to confirm validity of numerical results, we plan to use not only DuCOM-COM3 code but also ATENA code [9]. The results by the two codes are compared and checked. If we observe remarkable change in both results, we may proceed to more careful consideration. Based on this evaluation, repair, reinforcement, or other decisions should be taken into account. We are going to start practical

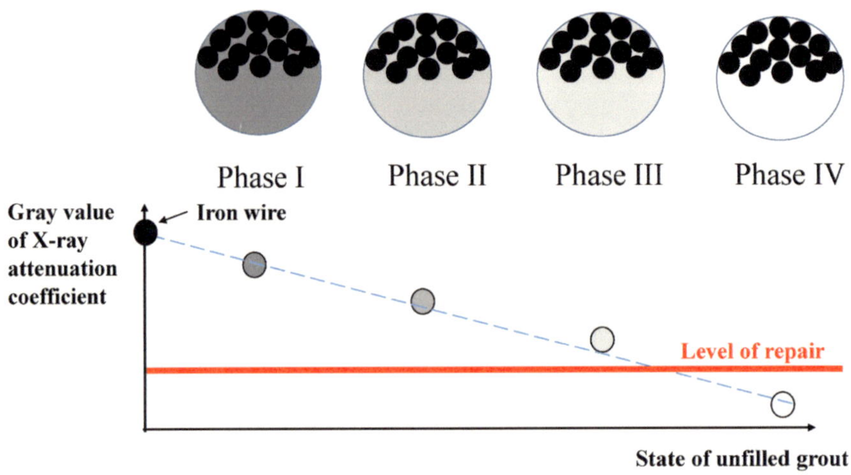

Figure 24.
Phases of unfilled grout and relation with gray value of X-ray attenuation coefficient.

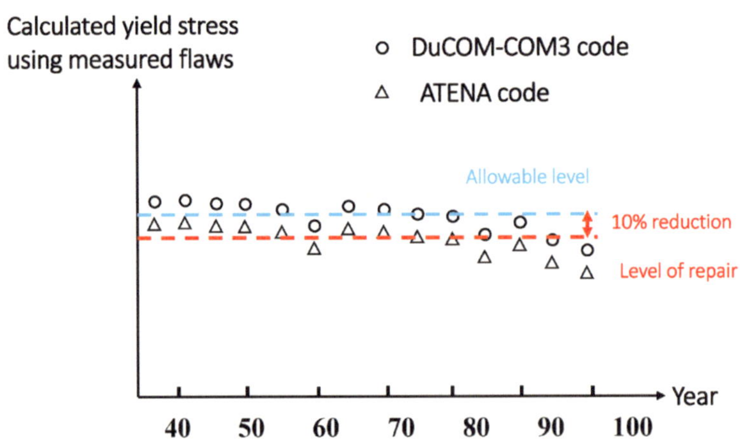

Figure 25.
Proposal of quantitative criterion for regular X-ray inspection and structural analysis with respective to measured property of mechanical strength.

and commercial highway bridge X-ray inspection under collaboration with major industries. We hope to apply these guidelines to all aged bridges in Japan and finally across the world in near future.

7. Conclusion

We are going to apply our 3.95 MeV X-ray source to highway PC bridge inspection in Japan in 2020. Among T-girder-, box- and slab-type bridges, we first inspect box type bridge with 550–750 mm thick WEB wall. We use 950 keV/3.95 MeV x-ray sources for thinner than 300 mm and 300–1,000 mm WEB walls, respectively. Main purpose is to visualize and evaluate unfilled grout in PC sheath. We performed preliminary experiments considering real situations to obtain X-ray transmission images of wires and unfilled grout of several PC sheaths. We successfully obtain one X-ray transmission image within 10 s. We also adopt boundary enhancement image processing to check the state of PC sheath, wires and grout. We also succeeded in obtain separate images of same vertical level PC sheaths by changing vertical location and angle of the X-ray source. In all cases, PC wires can been easily seen by black images and then we can recognize the location of PC sheath. Next, we check unfilled grout beneath the wires. We propose to use the gray value, namely relative X-ray attenuation coefficient, to evaluate of the state of unfilled grout.

Moreover, we try to apply 3.95 MeV neutron source in order to detect rainwater between surface asphalt and concrete. We successfully detected the existence of water via neutron back scattering in water. As a next step, we try to verify the rainwater detection in slab type bridge.

We also performed 3D structural analysis for box type bridge by using DuCOM-COM3 code. 5% reduction of yield stress due to thinning and disconnection PC wires via rainwater corrosion was evaluated. From now on, we are going to analyze mechanical strength reduction due to unfilled grout.

Finally, we propose a guideline for regular maintenance control by making use of 950 keV/3.95 MeV X-ray sources and structural analysis codes of DuCOM-COM3 and ATENA. After we verify the inspection of box type bridge in 2020, we plan to inspect other box- and T-girder-bridged in 2021.

Acknowledgements

This work was supported by the “Infrastructure Maintenance, Renovation, and Management program” of Cross-ministerial Strategic Innovation Promotion Program (SIP), Cabinet Office, Government of Japan in 2015-2019. We would like to thank Mr. Yasuhiro Ishida of the Public Works Research Institute of Japan for the total management, Prof. Yasuo Tanaka of Kanazawa Institute for Technology for the structural analysis and Mr. Hiroaki Takeuchi and Mr.Jean-Michel Bereder who are the graduates of Uesaka laboratory of Department of Nuclear Engineering and Management, the University of Tokyo for their related master thesis works.

Author details

Mitsuru Uesaka[1*], Katsuhiro Dobashi[1], Yuki Mitsuya[1], Jian Yang[1] and Joichi Kusano[2]

1 Nuclear Professional School, Institute of Engineering Innovation, The University of Tokyo, Ibaraki, Japan

2 Accuthera Inc., Kawasaki, Kanagawa, Japan

*Address all correspondence to: uesaka@tokai.t.u-tokyo.ac.jp

References

[1] Uesaka M et al. 950 keV, 3.95 MeV and 6 MeV X-band linacs for nondestructive evaluation and medicine. Nucl. Instrum. Methods Phys. Res. A. 2011;657:82-87.

[2] Uesaka M et al. Commissioning of portable 950 keV/3.95 MeV X-band linac X-ray source for on-site transmission testing. J. Adv. Maint. 2013;5:93-100.

[3] Uesaka M et al. On-site Non-destructive Inspection of the Actual Bridge using the 950 keV X-band Electron Linac X-ray Source. J. Disaster Res. 2017;12:578-584.

[4] Mitsuru Uesaka, Yuki Mitsuya, Katsuhiro Dobashi, Joichi Kusano, Eiji Yoshida, Yoshinobu Oshima, Masahiro Ishida, "On-site Bridge Inspection by 950 keV / 3.95 MeV Portable X-band Linac X-ray Sources", Bridge Optimization - Inspection and Condition Monitoring, IntechOpen, 2018, DOI: 10.5772/intechopen.82275

[5] Bereder J M. et al. Development of 3.95 MeV X-band linac-driven x-ray combined neutron source. Journal of Physics: Conference Series 874 (2017) (1) 012104. DOI: 10.1088/1742-6596/874/1/012104

[6] Pei C, Wu W, Uesaka M. Image enhancement for on-site X-ray nondestructive inspection of reinforced concrete structures. Journal of X-Ray Science and Technology. 2016;24: 797-805. DOI: 10.3233/XST, 160588s

[7] K. Maekawa, A. Pimanmas and H. Okamura: Nonlinear mechanics of reinforced concrete, Spon press, London, 2003.

[8] Robert GUINEZ Chef de l'unite technique Radiographie, Radioscopie Laboratoire Regional des Ponts et Chaussees de Blois and Eric HOYRUP, Directeur du Laboratoire Regional des Ponts et Chaussees de Blois Chef de l'agence Centre du CETE Normandie-Centre, Controle non destructive des ouvrages d'art par gammagraphie, radiographie et radioscopie, Bull. liaison Labo. P. et Ch. -171 -janv.-fevr. 1991 -Ref. 3550.

[9] Tereza Sajdlová, ATENA Program Documentation Part 4-8, ATENA Science – GiD Construction Process Tutorial, https://www.cervenka.cz/assets/files/atena-pdf/ATENA.

Chapter 4

Incremental Linear Switched Reluctance Actuator

Aymen Lachheb and Lilia El Amraoui

Abstract

Linear switched reluctance actuators are a focus of study for many applications because of their simple and robust electromagnetic structure, despite their lower thrust force density when compared with linear permanent magnet synchronous motors. This chapter deals with incremental linear actuator have switched reluctance structure. First, the different topologies of linear incremental actuators are mentioned. Furthermore, a special interest is focused on the switched reluctance linear actuator then the operating principal is explained. In addition, an analytical model of the proposed actuator is developed without taking account of the saturation in magnetic circuit. Finally, the control techniques that can be applied to the studied actuator are presented.

Keywords: linear actuator, switched reluctance, modeling, control, hybrid actuator, simulation

1. Introduction

Nowadays, linear actuators are used more and more in various industrial fields. This type of actuator makes it possible to have a direct linear drive without recourse to an intermediate motion transmission system [1]. Indeed, unlike conventional approaches where linear movement is obtained by coupling a rotary actuator to a movement transmission system, the direct generation of linear movement makes it possible to reduce the number of mechanical parts and therefore the losses associated with them.

The absence of motion transformers improves the overall performance of the system. As a result, the linear actuator is essential when the speed and precision are required by the application (machining Tools, manipulator robots, etc.), for this type of actuator the thrust force generated is thus applied directly to the load, [2]. Linear displacement controls are often used in industrial devices. In most cases where rigidity is required, it is provided by the worm and worm wheel system or the straight rack. These solutions introduce problems of transforming rotary motion into linear motion: slip, drop in efficiency, and bulk [3].

In some cases, a linear actuator may offer a satisfactory alternative when its construction and cost issues are resolved [4].

Robotic systems offer vast opportunities for actuators and among these, those with switched reluctance structure, rotary or linear.

The purpose of this study is to reflect the principle of the operation, and control of a linear actuator with the study of the different topologies of linear actuators.

IntechOpen

In the first part of this chapter the theory of switched reluctance machines well be explained. Then, the principle of operation as well as the analytical modeling of a linear actuator will be also studied. The last part of this chapter is devoted to the presentation of the control techniques dedicated for switched reluctance linear.

2. Classification of linear actuator

There are mainly three types of incremental linear actuator which differentiate by the physical phenomenon which is at the origin of their movement.

These three types of actuators can have structures with a planar or tubular geometry [5, 6]. Contrarily to rotating machines where the rotor and stator are generally coaxial. Linear machines can be presented in flat form or cylindrical form. They consist of a moving part and a fixed stator whose positions can be reversed.

For flat structures, it is possible to realize actuators with single stator or with double stator. For cylindrical structures, it is possible to consider tubular actuators with internal or external moving part.

The single stator actuator is a simple variance which is easily integrated in current applications but which presents a significant force of attraction between the stator and the moving part [7]. The double stator structure makes it possible to obtain, on the one hand, higher thrust forces than for the single stator structure and on the other hand to lighten the mobile part, because if the latter is well centered the resultant of the forces attraction is then zero. This structure is particularly well suited to the case where the fixed stator.

The linear actuator can also have two symmetrical inductors in order to create a greater force compared to its single inductor counterpart. Nevertheless, it has a complex geometry for its manufacture.

They are composed of a fixed part (the stator) and a mobile part (the translator) whose displacement is governed by the tendency of the magnetic circuits to be in a position of maximum flux.

2.1 Permanent magnet actuator

The permanent magnet linear actuator consists of an armature comprising one or more permanent magnets and a stator comprising a number of coils. There are two configurations of this type of actuator. The first is with fixed coils and moving magnets. The second is with moving coils and fixed magnets, **Figure 1** [6].

The operation of this type of actuator is provided by the action of an electromagnetic field on the armature made up of permanent magnets. The magnetic field in the air gap created by the supply of the phase coils orients the magnets in one direction.

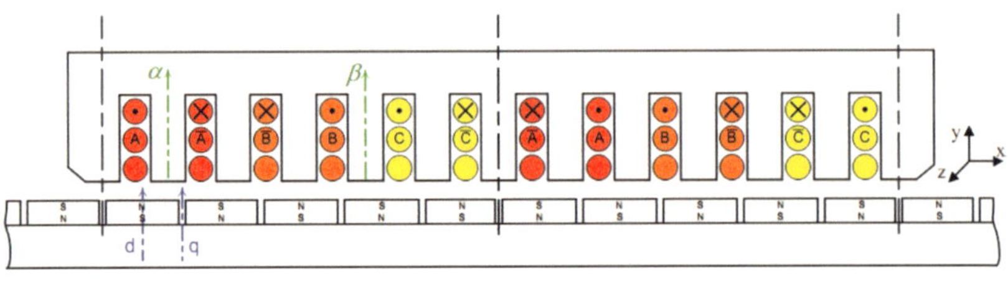

Figure 1.
Permanent magnet linear actuator.

2.2 Switched reluctance actuator

The switched reluctance actuator is among the simplest actuators. Regarding its construction, its basic structure consists of a coiled mobile part and an iron stator part which does not contain neither magnets or coils. The stator consists of an iron part rolled to form salient poles, **Figure 2**.

The principle of operation of a switched reluctance actuator is based on the tendency of an electromagnetic system to achieve a stable equilibrium position, which minimizes reluctance of the magnetic circuit. The aligned position of a phase is defined as the situation where the teeth of the stator and the modulus teeth of the mobile of the phase are perfectly aligned with each other reaching a position where reluctance is minimal [8].

Figure 3 shows an incremental reluctant linear actuator with transverse flux configuration comprising three modules separated by a non-magnetic material, each phase of the actuator is composed of two windings in series. The feeding of a phase creates a force allowing the movement of the mobile towards a stable equilibrium position, which it keeps as long as the power is maintained.

For this type of actuator, if the poles of one module are aligned with the poles of the stator then the poles of the other module must be offset in order to create a propelling force. Indeed, magnetic separations between the modules are necessary to impose a regular offset between the mobile modules.

2.3 Hybrid actuator

Hybrid stepping motors generally consist of a toothed mobile fitted with permanent magnets. The **Figure 4** shows the structure of a hybrid motor, [9].

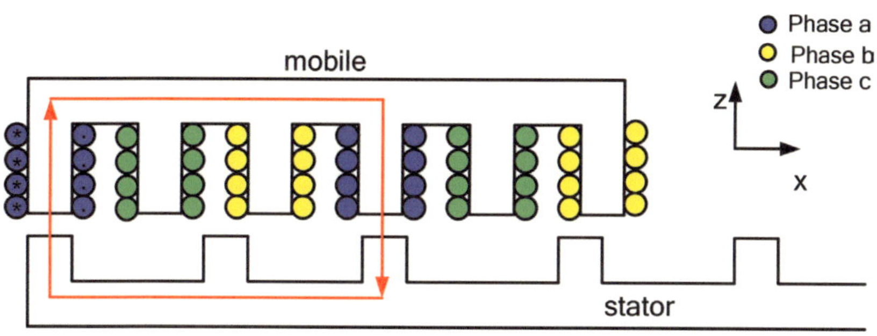

Figure 2.
Switched reluctance actuator with longitudinal flux configuration.

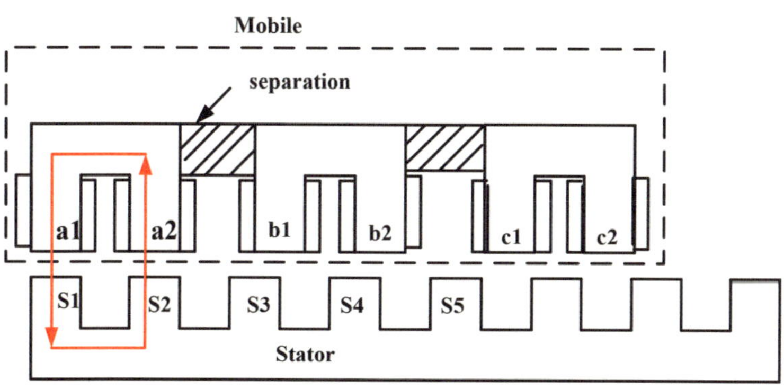

Figure 3.
Switched reluctance actuator with modular structure.

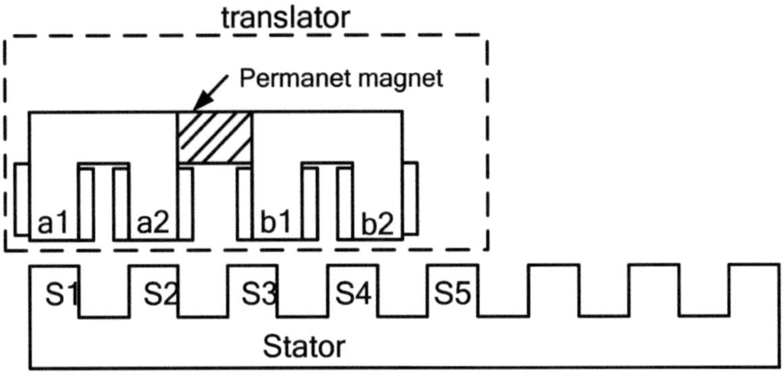

Figure 4.
Hybrid linear stepper actuator.

Type of actuator	Switched reluctance actuator	Permanent magnet actuator	Hybrid actuator
Resolution	High resolution	Medium	Medium
Thrust Force	Weak force	High force	High force
Operating frequency	high frequency	low frequency	high frequency

Table 1.
Comparison of different linear actuators.

This type of motor has both the advantages of the permanent magnet motor, which has a high torque, and those of the switched reluctance motor, which makes it possible to obtain a large number of steps per cycle. However, the iron losses are relatively large and therefore penalize this structure.

The movement of hybrid motors results from the superposition of the force developed by the reluctant effect of the teeth and the force created by the magnet.

The contribution of the amplitudes and the geometric periods of these forces makes it possible to achieve very diverse static characteristics. In fact, the magnet placed in the hybrid structure ensures a certain distribution of the field lines. The supply of the coils produces a switching phenomenon of the field lines more or less important depending on the intensity of the supply current by acting on the orientation of the fields it is possible to control the variation of the resulting force.

The **Table 1** gives a comparative study of the different configurations of the linear actuator studied above.

In what follows, our study will focus on switched reluctance actuators.

3. Operating principal of switched reluctance actuator

The force develop by a switched reluctance actuator is explained using the elementary principle of electromechanical energy conversion in a solenoid, as shown in **Figure 5**.

The switched reluctance machine belongs to the family of electromagnetic converters with single excitation.

Mechanical energy is produced by the displacement of a ferromagnetic material, placed in a magnetic field in order to maximize the flux in the circuit [10–13].

Generally, this type of actuator have only one degree of motion corresponding either to a translation or to a rotation around an axis, as shown in the figure.

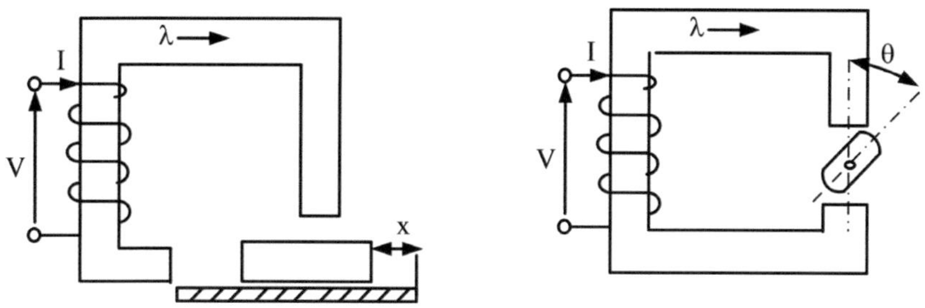

Figure 5.
Elementary circuit of a stepper motor.

The hatched parts represent the guiding of the moving part of which the position is identified by the distance x in the **Figure 5**. The position of the rotor in relation to the stator is indicated by the angle θ in the **Figure 5**.

Where ϕ and λ denote respectively the flow through a turn and the total flow of the winding.

$$\lambda = n\phi \tag{1}$$

For a given position of the moving part, the magnetic circuit is the seat of a totalized induction flux depending on the position and the supply current.

$$\lambda = \lambda(I, x) \tag{2}$$

To establish the equations governing the operation of the electromagnetic linear actuator, we consider the variations of the energy stored in the magnetic field when the moving part moves.

The axis of a moving tooth is identified by its distance x with the axis of a fixed notch. Thus, we define:

$$\text{Magnetic energy}: W_{mag} = \int I d\lambda \Big|_{I=cte} \tag{3}$$

$$\text{Magnetic co-energy}: W_{co} - \int \lambda dI \Big|_{\lambda=cte} \tag{4}$$

4. Analytic modeling of linear switched reluctance

Investigating the operational behavior of the switched reluctance actuator requires a mathematical model based on the electrical and mechanical equations governing its operation.

4.1 The electrical model

An elementary equivalent circuit for the switched reluctance actuator can be obtained by neglecting the mutual inductance between the phases. Assuming that each phase of the motor consists of a coil with resistance R and inductance $L(I, x)$, the applied voltage U to a phase is equal to the sum of the resistive voltage drop and the derivative of flux linkages λ (I,x).

$$U_j = R_j I_j + \frac{d\lambda_j(x, I)}{dt} \tag{5}$$

j represents the index of the phase with j = 1, 2,3, indicates in order the phases A, B and C.

The flux linkage depends on the current and the translator position. Then the flux expression becomes:

$$U_j = R_j I_j + \frac{\partial \lambda}{\partial I}\frac{dI_j}{dt} + I_j \frac{\partial \lambda_j}{\partial x}\frac{dx}{dt} \quad (6)$$

By mean the magnetization curve **Figure 6**, the magnetic field energy can be determined for a fixed translator position as function of current and linkage flux.

$$W_{co} = \int_0^I \lambda dI \quad (7)$$

The force developed by the switched reluctance linear actuator is proportional to the change in mechanical energy as a function of mechanical displacement. It can be given by:

$$F(I,x) = \frac{\partial W_{co}}{\partial x} \quad (8)$$

The total instantaneous electromagnetic force Ft is the sum of the q individual phase forces.

$$F_t(I,x) = \sum_{j=1}^{q} F(I,x) \quad (9)$$

For a linear flux model of switched reluctance actuator it is $\lambda(x) = L(x)I$. Thus, neglecting magnetic saturation gives.

$$W_{co} = \int_0^I \lambda dI = \frac{1}{2}L(x)I^2 \quad (10)$$

Hence,

$$F(I,x) = \frac{1}{2}I^2\frac{dL}{dt} \quad (11)$$

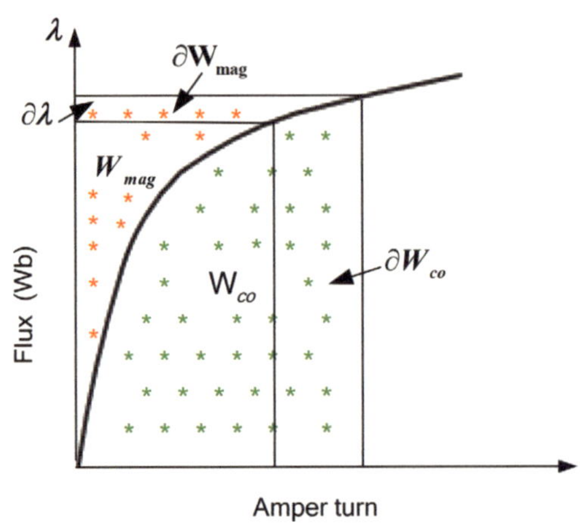

Figure 6.
Magnetization curve of the studied actuator.

The thrust force is then proportional to the derivative of the inductance with respect to the displacement of the mobile x and to the square of the supply current of a phase.

The inductance $L\ (I,\ x)$ is a periodic function of x, with a period equal to the dental pole pitch δ .

The inductance is minimal when the teeth are in unaligned position, and it is maximum when the teeth are in aligned position.

The inductance depends on the moving position. The partial derivative of the inductance with respect to the moving position can be expressed by:

$$\frac{\partial L_j}{\partial x} = -\frac{2\pi}{\delta} L_1 \sin\left[\frac{2\pi x}{\delta} - (j-1)\frac{2\pi}{3}\right] \tag{12}$$

By replacing Eq. (12) in Eq. (11), the electromagnetic force developed by each phase of the separately supplied switched reluctance actuator is expressed as follows:

$$F_j(x,I) = \frac{1}{2} I_j^2 L_1 \frac{2\pi}{\delta} \sin\left(\frac{2\pi}{\delta} x - (j-1)\frac{2\pi}{3}\right) \tag{13}$$

4.2 The mechanical model

The mechanical movement of the actuator is described by the equation deduced from the fundamental principle of dynamics characterizing a linear movement, (see Eq. (14)) [5].

$$m\frac{d^2 x}{dt^2} + D_v \frac{dx}{dt} + f_s \text{sing}(v) + F_l = F(x,I) \tag{14}$$

5. Specificity of the control of linear actuator

The special feature of the switched reluctance linear actuator is to ensure a continuous incremental translational movement. In other words, each supply pulse must correspond to a constant elementary displacement, this is correspond to the mechanical step of the actuator.

Then, to ensure continuous movement, it is necessary on the one hand to have several phases, on the other hand, the successive supply of the phases must be synchronized with the real position of the moving part.

A determined number of pulses causes a corresponding number of steps by the actuator. In addition, the succession of determined pulses generated by a control circuit at a well determined frequency makes it possible to impose a continuous movement of the moving part at constant speed. At each pulse of the control, induce that the poles of the supplied phase closest to the stator poles, and they are positioned opposite the latter.

Like all electric motors Linear, actuators can be driven in open loop or closed loop for applications, which require high precision and high positioning quality.

The linear actuator is essentially an electric actuator, which requires an electronic power converter to change the operating frequency and the magnitude of the applied voltage. The main characteristics of electronic converters used for linear actuators generally require full operation of two quadrants (half H-bridge), a high switching voltage is necessary for the rapid establishment and extinction of the current.

Generally the linear incremental actuators are controlled by a static converter in most applications, the force generated by each phase of the actuator is proportional to the square of the phase current (see Eq. (13)). The circuit and the control strategy are directly related to the performance and characteristics of the actuator. Several topologies are presented with a reduced number of power switches, faster excitation, faster demagnetization, high efficiency and high power through continuous research [14]. Conventionally, there has always been a trade-off between obtaining some advantages and losing others with each topology.

Assuming that the edge effects is neglect, then the variation of inductance is linear as shown in **Figure 7**. The characteristic of the inductance is periodic and the periodicity of the inductance is equal to $2\pi / q$, q number of phases.

The physical meaning of the different regions in **Figure 7** is as follows.

In the [*x1-x2*] zone, the inductance begins to increase as the mobile moves. When the poles of the moving part meet the stator poles, the inductance reaches its maximum value. In this region, the actuator operates in an increasing inductance regime where the slope of the inductance is positive where a positive force is developed by the actuator.

The area in the gap [*x2-x3*] the teeth of the mobile and the stator are completely aligned. In this interval, the inductance is constant and in this case the actuator cannot generate any force even if the phase supply is kept constant.

5.1 Different control methods

Incremental actuator supplies are generally classified into five modes [14]:

Mode 1: only one phase is supplied by the nominal current I_n, in this case that the mechanical step of the actuator is defined, the phase supply sequences is shown in **Figure 8**.

Mode 2: two successive phases are supplied at the same time by the current. Indeed, the force is greater by a factor $\sqrt{2}$ than the first mode.

Mode 3: the alternating combination of the two previous modes allows operation in half-step, In this control mode, the phases of the actuator are supplied in order in accordance with the cyclogram shown in **Figure 9**.

Mode 4: this mode, commonly called "Ministepping" consists in multiplying the intermediate positions by supplying each phase with fractions of the nominal current, this corresponds to the extension of operation in mode 4.

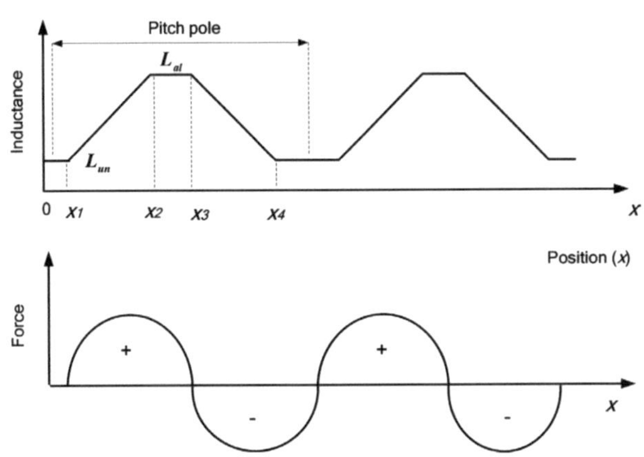

Figure 7.
Inductance and force curves as function of position.

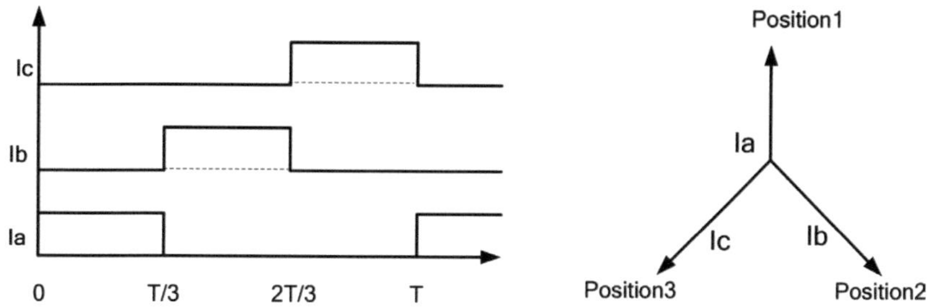

Figure 8.
Full step command.

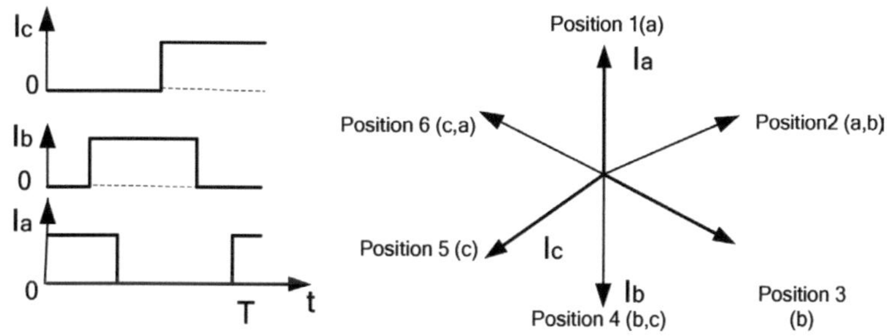

Figure 9.
Half steps command.

5.2 Drive circuit for switched reluctance actuator

Several application using linear switched Reluctance actuator using an adjustable speed drive. Asymmetric half bridge converter is very popular for switched reluctance actuators because of its ability to operate efficiently.

The power switch used in converter is a transistor. However, in industrial applications, the other types of power switches are used, mostly Thyristor, power IGBTs, or even MOSFETs. A dc voltage source is necessary for supplying the power converter and motor phases. The dc source may be from batteries or mostly a rectified ac supply with a filter to provide a dc input voltage source to the switched reluctance actuator converters (**Figure 10**) [14].

First, consider that the phase *1* is supplied when the upper and lower transistors *T1* and *T4* are switched on. Then, + V_{DC} voltage applied to the phase winding. Therefore, a current is established and increases in the windings of the phase through both switches.

When the poles of the supplied phase reach the aligned position with the poles of the stator, in this case the switches are turned off. Phase current then slowly decreases by freewheeling through one transistor and one diode. When both transistors are off, the phase winding will supplied by voltage -V_{DC}. Indeed, the phase current then quickly decreases through both diodes. By appropriately coordinating the above three switching states, phase current of the switched reluctance actuator can be controlled. The major advantages of the asymmetric bridge converter are the independent control of each motor phase and the relatively low voltage rating of the inverter components.

By supplying the first phase maintains the translator in a stable equilibrium position, if the supply current of the first phase is cut off and if the second phase is supplied, then a positive force is developed by the second phase which moves the translator to the second equilibrium position. Conversely, if the excitation is changed from phase *1* to phase *3* the force developed by phase 3 is negative, moving the translator in the negative direction to the phase *3* equilibrium position.

The **Figure 11** presents the characteristics of the forces developed by a three-phase structure supplied separately. By applying a resistant force F_l when phase 1

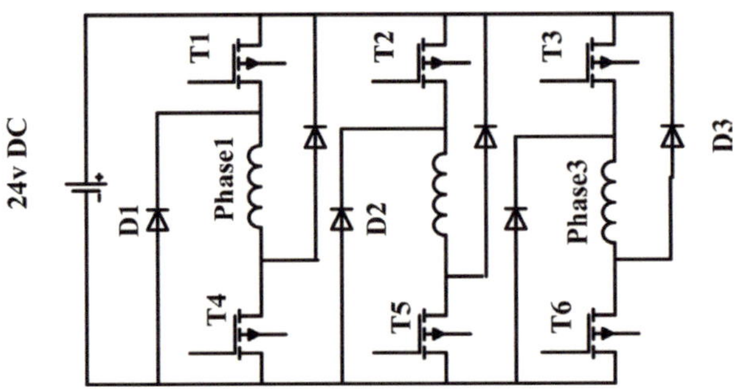

Figure 10.
A three-phase asymmetric half-bridge converter.

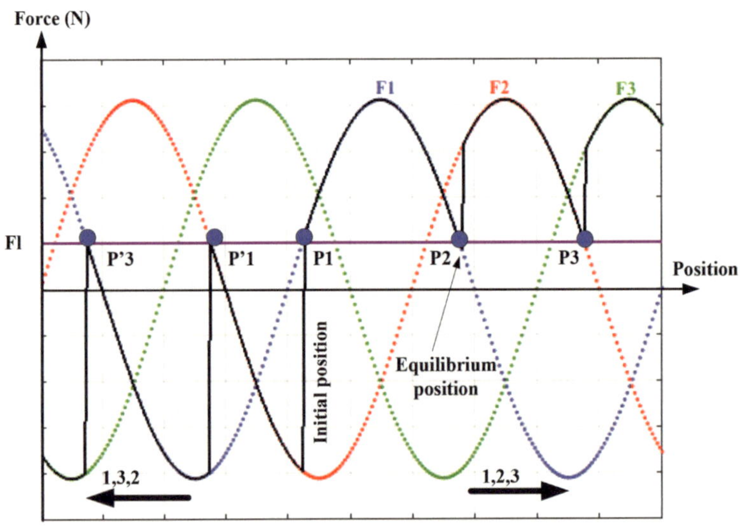

Figure 11.
Force characteristic of an actuator possessed three phase.

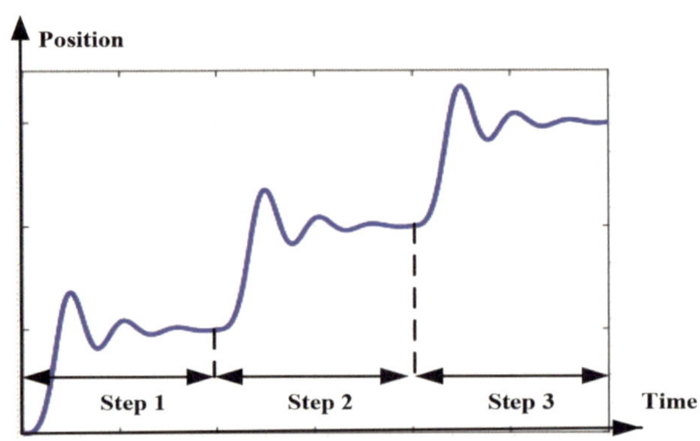

Figure 12.
Dynamic response of the actuator for full step command.

is supplied, the stable equilibrium position is at P1. If the second phase is powered, the mobile is brought to position P2 when phase 3 is powered in turn, it brings the mobile to P3 and so on. The power supply sequence in the order 1,3,2 makes it possible to move the mobile in the opposite direction from P1 to P′1 then to P’3.

The **Figure 12** show the response of the actuator when a full step command is applied. The successive feeding of the phases causing the moving of the mobile by three mechanical steps, we can see that at each step the mobile oscillates around its equilibrium position in order to reach a stable position.

6. Conclusions

In this chapter, initially, the different topologies of incremental linear actuators are studied and presented, operating principal of switched reluctance actuator was described. Then, an analytical model of linear switched reluctance actuator was proposed and established which allowed to determine the electromagnetic force developed by the actuator. Finally, the different control techniques that can be applied to the studied actuator have been presented.

Conflict of interest

The authors declare no conflict of interest.

Nomenclature

U	Voltage (V)
I	current (A)
R	phase resistor (Ù)
L	inductance (mH)
λ	Linkage flux (wb)
δ	Pole pitch
W_{co}	Co-energy (J)
m	mass of the moving part (Kg)
D_v	viscous coefficient of friction (Ns/m)
fs (N)	dry friction
v (m / s)	represents the mechanical speed of the mobile (m/s)
x	the displacement (m)
Fl	the load force (N)
F	the electromagnetic force (N)

Author details

Aymen Lachheb* and Lilia El Amraoui
Research Laboratory Smart Electricity and ICT SEICT Lab LR18ES44, National Engineering School of Carthage, University of Carthage, Tunisia

*Address all correspondence to: aymen.lachheb@enicarthage.rnu.tn

References

[1] Antonio Eduardo Vitoria, et al: Design and evaluation of a linear switched reluctance actuator for positioning tasks. *Turkish Journal of Electrical Engineering and Computer Science*. 2010: 925-941. DOI: 10.3906/elk-0911-295.

[2] Dursun M, et al: Design of Linear Switched Reluctance Motor Driver for Automatic Door Application. International Journal of Information and Electronics Engineering 2013; 237-241. DOI: 10.7763/IJIEE.2013.V3.307.

[3] Lachheb A, El Amraoui L and Khediri J: Finite Elements Modeling of Linear Motor for Automatic Sliding Door Application. International Journal of Advanced Computer Science and Applications 2016; 106-112.DOI: 10.14569/IJACSA.2016.070816.

[4] Lachheb A, Khediri J and El Amraoui L: Performances Analysis of a Linear Motor for Sliding Door Application. *International Journal of Power Electronics and Drive Systems* 2017; 1139-1146. DOI: 10.11591/ijpeds.v8.i3.pp1139-1146.

[5] Marcel J, Electromécanique presses Polytechniques et Universitaires romandes; 1995.

[6] Jacek F, Zbigniew J and Bronislaw Z: Linear Synchronous Motors. *2nd Edition Tayloy & Francis.*

[7] Jordi Garcia A, Pere A and Baldui B. Linear switched reluctance Motors Modeling and Control of Switched Reluctance Machines Publisher: *Interchopen* 2020. DOI: 10.5772/intechopen.89166.

[8] Zaafrane W, Rehaoulia H, X Dursun M, et al: Low cost linear switched reluctance motor velocity and position control. Journal of Circuits Systems, and Computers, 2016: 1650021-1650021. https://doi.org/10.1142/S0218126616500213.

[9] J.Amorós1 et al Two-Phase Linear Hybrid Reluctance Actuator with Low Detent Force: *Energies, 2020:* DOI: 10.3390/en13195162

[10] krishnan R, Switched reluctance motor drives CRC Press 2001.

[11] B. Seok L. Linear Switched Reluctance Machine Drives with Electromagnetic Levitation and Guidance Systems [thesis]. Blacksburg, Virginia, 2000.

[12] J. Amorós, P. Andradab and B. Blanqué: Design Procedure for a Longitudinal Flux Flat Linear Switched Reluctance Motor. *Electric Power Components and Systems*, 2011: 161-178. DOI: 10.1080/15325008.2011.629333

[13] Fenercioglu A and Avsar Y: Design and analysis of EI core structured transverse flux linear reluctance actuator. Turkish Journal of Electrical Engineering and Computer Sciences. *2013;* 945-955. DOI: 10.3906/elk-1304-145.

[14] Acarnley P, Stepping Motors a Guide to Theory and Practice *The Institution of Electrical Engineers*, London: Edition IEE, 2002.

Chapter 5

A Survey on Weapon Target Allocation Models and Applications

Abdolreza Asadi Ghanbari, Mousa Mohammadnia, S. Abbas Sadatinejad and Hossein Alaei

Abstract

In Command and Control (C2), Threat Evaluation (TE) and Weapon Target Allocation (WTA) are two key components. To build an automated system in this area after modeling Threat Evaluation and Weapon Target Allocation processes, solving these models and finding the optimal solution are further important issues. This setting demands instantaneous operational planning and decision making under inherent severe stress conditions. The associated responsibilities are usually divided among a number of operators and also computerized decision support systems that aid these operators during the decision making process. In this Chapter, the literature in the area of WTA system with the emphasis on the modeling and solving methods are surveyed.

Keywords: command and control (C2), weapon target allocation (WTA), mathematical models, algorithmic approaches, decision support systems (DSS)

1. Introduction

The field of air defense is one of those areas where resource allocation is of great importance. Research on the resource allocation problem with military purposes dates back to the 1950s and 1960s where the first modeling issues for WTA problem were investigated [1]. Rapid developments in the field of battle and attention to advances in threat technology in the recent years, pose significant challenges for commanders in C2 Systems (CCSs). Furthermore, the complexity and diversity of engagement scenarios, and the volume and imperfect nature of data to be processed under time-critical conditions make the commander's problems severe.

The WTA problem is a well-known military operations research problem which has many aspects and features. (a) It is a dynamic decision-making problem. Serial and interdependent decisions are made at different periods of time to deal with different threats. At each period, a decision (about one or several engagements) is made. The consequences or outcomes of decisions made at a given period change the characteristics of the problem for the next periods (e.g. ammunition availability, threats conditions, change in tactics, decreasing the data ambiguity). One important characteristic of dynamic decision-making problems is that the information is obtained gradually over time. (b) The WTA is a multiple criteria decision-making problem. In fact, the decision-making problem for the resource management is based on conflicting criteria. For example, minimizing the risk, maximizing the effectiveness measures and so on. (c) The resource allocation problem is subject to uncertainty (e.g., stochastic). The uncertainty is related to many aspects of the model such as the hit probability of the weapons and targets characteristics (e.g. maneuvering, tracking, identification/classification and so on). Therefore, decision consequences are uncertain and usually modeled by probabilistic distributions. Then, a kill assessment process checks the result of executed actions and reports either the threat is diminished or not. (d) The resource allocation problem is a time-critical decision-making problem. Decision makers have to decide and act under tight temporal constraints. The allocation of available weapons on priority bases to the correct targets needs complex calculations in a very short time.

Some important properties of the WTA problem are:

- It is NP-complete and consequently the computation time of any optimal algorithm grows exponentially with its size [2]. The complexity of this problem drastically increases if the temporal and spatial constraints of both the Blue Forces (BF) (i.e. friendly forces) and Red Forces (RF) (i.e. hostile targets) are considered.

- It is sequential (the results of previous engagements are observed before making present assignments) [3].

- The objective function is nonlinear (The WTA objective, often is a linear combination of nonlinear summand of expressions like $[1 - P_{ik}]^{X_{ik}}$) [4].

- It is stochastic (weapon-target engagements are modeled as stochastic events) [5].

- It is large-scale (the combination of weapons-targets-time, grows exponentially) [2].

- It is mixed-integer; WTA is a combination of discrete and continuous variables.

Finding efficient solutions for WTA problem with these features, is the main challenge of the command. These WTA characteristics often make the solving process and finding the optimal solution difficult.

Researchers have suggested different mathematical formulations of the WTA problem and proposed exact algorithms and heuristic/meta-heuristic methods to solve the problem. The exact algorithms that have been introduced to solve the WTA problem are not comprehensive and usually run under the following conditions: (i) when all the defensive weapons are identical [6] or (ii) when the hostile targets can receive at most one defensive weapon [7]. Despite extensive attention of research in this field, a comprehensive and classificatory review will pave the path for more efficient and practical models and algorithms in accordance with the advent in the technology. Hence, in this survey different WTA models and solutions methods used for this problem are reviewed. Based on the modeling approaches, the available literature is classified into two groups. (i) defense allocation models, (ii) the game models. In the second part, the important characteristics for the WTA models and the solution process for the resource allocation problem are investigated. Like many other problems the WTA solution algorithms can be divided into three main categories, (a) enumerative techniques, (b) heuristic/approximate methods and (c) meta-heuristic methods. Finally, mission planning systems which have been developed in the military are listed. To cover the existing shortcomings and improve the weakness of models and solutions procedures some useful suggestions have been put forward for future studies.

The organization of this Chapter is as follows. In Section 2, we give a brief description of the problem and its components. In Section 3, the basic formulations of the WTA problem are presented and variations of modeling approaches are explained. Section 4 contains classification for algorithmic approaches. The last section contains some concluding remarks and providing directions for future research.

2. The WTA problem and its components

There are many different functions that must be handled in an air defense system. From a high-level command perspective, these functions can be divided in two main processes: picture compilation and picture exploitation [2].

The purpose of picture compilation process is generating a representation of the volume of interest (VOI) based on data and information received from a variety of sensors/sources. This step often includes the following sub-processes: (i) The target detection, (ii) target tracking, and (iii) target identification. The picture exploitation is the subsequent process that includes the following sub-processes: (i) Threat Evaluation (TE), (ii) Engageability assessment and (iii) weapons assignment (here referred to as WTA) [2]. In a military context this set of sub-problems also are known as combat power management (CPM). The WTA component, in a CPM system can be divided into three sub-problems:

- *Response planning*; this sub-problem deals with the combat resource allocation. During this process, one or more of available weapons are assigned to engage each threat.

- *Response execution*; in this phase the planned response is executed at the designated time.

- *Outcome assessment*; in this process the outcome of the executed actions or engagement is identified and/or verified. This process is well known as killing assessment (where outcomes are 0/1 for kill/alive) or damage assessment (where outcome is a value in [0, 1] i.e. partial damage is possible) in C2 domain.

This research concentrates only on the response planning procedure of WTA process. This process is about how and when to allocate the available defensive resources. In other words, response planning procedure has two aspects: resource allocation planning and scheduling.

2.1 Resource allocation: planning and scheduling

In C2 context the response planning mainly is pertained to resource allocation problem. Resource allocation is the assignment of resources to activities, where the start and end times of each activity is given [8]. This problem is concerned with optimally assigning weapons to the hostile targets so that after all engagements, the BF expectations are met as far as possible. Resource allocation scheduling is another important aspect of the response planning that consists of determining the start and end times of the activities. In pure scheduling problems, activities are already chosen (or given), leaving only the problem of determining a feasible and possibly best order among them. In defensive operations, scheduling determines when a specific defensive action (e.g. assigning a specific weapon against a specific threat) ought to take place [9]. Therefore, the response planning determines the best assignment of the resources and specifies the start and end times of the activities. This procedure defines a joint resource allocation planning & scheduling problem. In this Chapter we focus on the response allocation planning or determining optimal assignment of weapons to the hostile targets which is referred as WTA.

It is interesting to note that, in contrast with other CPM components like TE, WTA is well documented in the literature and possible solutions have been investigated for a number of years. The WTA process is one of critical decision making, that is consistent with the related own force mission objectives and compatible with the Rules of Engagement (RoE), Weapon Systems (WSs) and environmental constraints. In the next section the basic formulations of the WTA problem and variations of modeling approaches are explained.

3. WTA modeling approaches

The WTA problem is a model of combat operations where a finite number of weapons are assigned to hostile targets (i.e. RF). The BF expectation and objective must be fulfilled after all engagements. The BF objective could be minimizing the total expected value of surviving targets, minimizing military resource costs and so on. From mathematical point of view, the purpose of the WTA solving process has been determined as weapon-target pairs. In the following, this process will be described formally, but first, we need to introduce some notations. The notations employed in this context are listed in **Table 1**.

3.1 WTA basic models

Assuming a defensive scenario consisting of $|W|$ firing units and $|T|$ targets, from a defensive perspective, there are three basic models for WTA [5].

Basic model 1: The first basic model for WTA problem, which focuses on maximizing damage to the enemy/minimizing total expected target values, can be formulated as:

$$\min \; F = \sum_{i=1}^{|T|} V_i \prod_{k=1}^{|W|} (1 - P_{ik})^{X_{ik}} \tag{1}$$

In this model the goal of the defender is to minimize the numbers of the surviving invading enemy, using the optimal allocation of available weapons.

Basic model 2: The second basic model for WTA is to allocate the available firing units to maximize the total expected protection value of surviving defended assets. This model can be formulated as follows:

$$\max \quad J = \sum_{j=1}^{|A|} \omega_j \prod_{i \in G_j} (1 - \pi_{ij} \prod_{k=1}^{|W|} (1 - P_{ik})^{X_{ik}}) \tag{2}$$

Basic model 3: If the total expected value of surviving assets after the current stage is employed as the objective of WTA for the current stage, then the formulation of the objective function for basic Dynamic WTA in stage t is well be as follows [10]:

Sets
T_i: the set of detected threats, $i = 1, 2, \cdots, I$.
w_k: the set of resources, $k = 1, 2, \cdots, K$.
A_j: the set of assets $j = 1, 2, \cdots, J$.
$s = 1, 2, \cdots, S$: the set of engagement stages.
Parameters
P_{ik}: the estimated effectiveness, i.e. probability that resource $w_k \in W$ neutralize threat $T_i \in T$ if assigned to it.
π_{ij}: the estimated probability that threat $T_i \in T$ destroys the asset $A_j \in A$.
V_{ij}: the threat value of the threat-asset pair (T_i, A_j).
ω_j: the protection value of asset $A_j \in A$.
C_{ik}: a resource usage cost for assigning resource $w_k \in W$ to target $T_i \in T$.
Variables
$X_{ik} = \begin{cases} 1 & \textit{if resource } w_k \in W \textit{ is assigned to target } T_i \in T \\ 0 & \textit{otherwise} \end{cases}$
$[X_{ik}^s]_{I \times K}$: is a decision matrix at stage s.

Table 1.
Notations.

$$\max \quad J_t(X^t) = \sum_{j=1}^{|A(t)|} \omega_j \prod_{i=1}^{|T(t)|} \left[1 - \pi_{ij} \prod_{h=t}^{S} \prod_{k=1}^{|W(t)|} \left(1 - p_{ik}(h)\right)^{x_{ik}(h)} \right] \quad \text{with } t \in \{1, 2, \ldots, S\} \tag{3}$$

Where h is an index of stages and the defender have S engagement opportunities which depends on the time–space conditions of the interception and weapon characteristics. Here $A(t)$, $T(t)$ and $W(t)$ are sets of defended assets, hostile targets and available weapons in stage t, respectively. These three basic models are reference for the other models which have been incorporated in the applications.

3.2 Variation of modeling approaches

The WTA problem may be considered from a number of different perspectives. In this section, first a framework is provided for identifying key elements of WTA problem and then based on these characteristics we describe particular sets of modeling techniques which have been proposed for these problems.

In terms of interaction with the environment and modeling of these interactions, there are two general formulations that are investigated in literature for WTA problem: static and dynamic WTA model.

1. *Static WTA (SWTA),* in this approach of modeling, all weapons engage targets in a single stage and assume that the decision maker is aware of all parameters for the problem, while the outcomes of the assignments could be stochastic. Thus, the goal of SWTA is to find the optimal assignment for a temporary defense task (the basic models 1 and 2 are static models). This WTA problem was first formally posed in 1958 [1]. It is interesting to note that the majority of the literature in WTA domain have been dedicated to efficiently solving the SWTA problem formulation and few optimal methods have been presented for

solving SWTA with some simplifications. For example, optimal methods have been proposed by denBroeder [11] under a homogenous weapon set assumption, known as the maximum marginal return (MMR) algorithm. Orlin [12] developed optimal algorithms under the assumption that each target can have no more than one weapon assigned to it. A survey of WTA solution methods has been provided in section 7 of present Chapter.

2. *Dynamic WTA (DWTA),* in this approach, WTA is molded as a multi-stage problem and the outcome of each engagement is assessed for subsequent decisions. This variation of WTA problem was firstly formulated by Hosein et al. in 1989 [13]. The DWTA problem has similar stochastic elements as the SWTA problem, but assignments are made in multiple stages. In other words, the time parameter is defined in dynamic models and the goal is to find a global optimal assignment for the whole defense process. The DWTA often have been presented as a sequential decision making problem. The basic assumptions of the DWTA are outlined as follows [14]:

    - In each stage, a subset of the available weapons is selected and allocated simultaneously.

    - The outcomes of each stage are observed prior to the preceding stage (i.e. the problem re-solved at each stage using previous stage information).

The DWTA problem suffers from the curse of dimensionality. Therefore, traditional solution methods become intractable. The curse of dimensionality arises from three elements of the problem i.e. state space, decision space, outcome space, or their combination [2]. Intuitively, DWTA can be achieved by a series of SWTAs through all stages. However, although SWTA can at best, guarantee the optimality of the WTA decisions for its corresponding defense stage, the combination of all SWTA decisions may not be optimal for the whole defense process (because of its greedy nature). In a real-world situation, the decision of which weapon system to be allocated to which target needs to be complemented with scheduling of when to engage a target.

Moreover, apart from the modeling approaches being static or dynamic; the WTA problem can be formulated based on two different defense scenarios. The air defense capability may be limited to self-defense or may be extended to area defense where a set of assets within a certain range should be defended. For example, in a military task group, the individual defense units work together as a team to provide mutual support and defense against incoming threats. These units are typically arrayed into a formation, called a screen, in which the most valuable and important units (i.e. high value units (HVUs)) are surrounded and protected by the escorting units (e.g. vessels and airplanes). Within the screen, the escort units often are stationed in sectors away from the HVU. In such a situation important issues such as positioning and coordination between these units are aroused [10]. Therefore, in WTA problem the following two perspectives are usually adopted:

1. *The single platform perspective*: a single platform intends to protect an asset (may be itself) from incoming threats, where assignment relates to selecting the most suitable WS to counter a threat (e.g. a ship that intends to defend itself against invading threats). In this case often the defense and the defended property (i.e. assets) are the same.

2. *The force coordination perspective:* a C2 platform providing protection (e.g. TE and WTA) for third party Defended Assets (DAs), where assessment relates to

the identification of the most suitable armed platform to engage or counter a threat. In this case the defense and the defended property are different (likely separated) and the purpose of defense is to maximize the chance of survival of the being-defended assets.

The main objective of the invading enemy is harming the defended valuable assets, but for making these assets defenseless, they may first destroy the protective weapon platforms. So the arrangement of platforms (i.e. defense units) is one of the major issues that must be handled in this type of WTA problem. This positioning must maximize asset protection while minimizing the possible damage to them [10].

In addition to the number of deployed defensive units, the defense scenarios usually involve more than one hostile target. Therefore, when investigating the WTA modeling approach where the available WSs are assigned, two perspectives are prevalent:

1. *The threat-by-threat perspective,* which refers to the assignment of WSs sequentially in such a way that the best WS is essentially assigned to each threat in turn (from the highest priority to the lowest priority).

2. *The multi-threat perspective,* which refers to the assignment of WSs to the current set of threats concurrently so that the assignment is best in some overall sense.

Technically, both of the above mentioned approaches employ some form of optimization techniques. What distinguishes them is the threat-by-threat assignment which usually is based on some type of greedy algorithms, whereas the multi-threat assignment typically involves the optimization of a given objective function (e.g. maximize the damage to the targets/minimize the value of any remaining targets, maximize the asset protection and so on) subject to certain constraints (e.g. the target priority, the number of simultaneous engagement and so on) [10].

Furthermore, depending on the defense priorities and attitude of the modeling, the WTA problem can be stated as two different optimization problems:

1. *Target-based (weighted subtractive) defense;* the BF allocates firing units to minimize the expected total target value of the RF targets.

2. *Asset-based (preferential) defense;* the WTA problem is stated as an optimization problem where the objective becomes to maximize the total expected survivability of the defended assets.

Target-based objectives lead to subtractive defense strategies (basic model 1). In other words, the defense tries to destroy as many of the most lethal targets as possible, or at least the most valuable ones. On the other hand, asset-based objectives lead to preferential defense strategies (basic models 2 for SWTA and basic models 3 for DWTA). In order to save an asset, the defense must, with high probability, destroy all of the targets aimed for it or divert them. For these purposes each of the hostile targets must be attacked with appropriate hard-kill or soft-kill weapons (e.g. jammers, decoys and etc.). In this case the defense may not have enough weapons to defend all the assets. Therefore, the defense must decide which of the assets should be protected (i.e. which one is preferred) and to assign its defensive resources to protect them. However, no weapons may be assigned to the targets aimed for the other low valued defended assets (i.e. this asset is immolated to save other valuable assets). This is known as a preferential defense strategy.

The main difference between the target-based WTA problem and the asset-based WTA problem is that in the latter, the aims of the enemy are assumed to be known. In other words, the asset-based formulation demands knowledge of which hostile targets are headed for which defended assets. Therefore, the asset-based WTA problem formulation is suitable for ballistic missile defense problems (such as base camp protection against artillery and mortars) while the target-based formulation is more appropriate when the intended aims of the targets are not known for sure [15]. The target intention determination can be thought of as a data association problem [2].

In defense scenarios, the defender may look for optimizing different objectives. According to the number of goals which need to be optimized, the WTA is modeled in two ways:

1. *Single-objective WTA problem*: if the quantity to be optimized is expressed using only one objective function, the WTA problem is modeled as a single-objective problem. The objective function might be minimizing survival probability of hostile target or maximizing protection value of surviving defended assets (see e.g. Eq. 1 and 2).

2. *Multi-objective WTA problem*: if the problem concerns on many objectives simultaneously then WTA will be a multi-objective problem. These objectives cloud be minimizing the risk and cost, maximizing the effectiveness measures and so on. Normally these objectives are conflicting each other.

The WTA problem has been designed to support human operators in real world condition. Therefore, the proposed WTA model should be able to cope with numerous and heterogeneous real-world objectives and constraints to help the operator to make appropriate decisions. This implies that WTA is required to be formulated as a multi-objective optimization (MOO) problem. However, relatively few researchers studied the WTA as a MOO problem. The research work by Ghanbari et al. [3], Newman et al. [16] and Leboucher et al. [17] applied MOO to formulate the model for the WTA problem.

A summarizing list of WTA modeling approaches from a BF perspective is given in **Table 2**. In WTA modeling miscellaneous limitations must be considered, these constraints will be discussed in the next section.

From **Table 2** it can be deduced that the simplest approach for formulation the WTA problem is a static, single-objective, target-based, threat-by-threat and single platform model; while the most complex formulation is a dynamic, multi-objectives, asset-based, multi-threat and force coordination model. The asset-based formulation increases the model complicity because it demands hostile targets intention analysis. The models proposed in WTA literature are in a state between the easiest and the hardest one. While the real-world situations tend towards to the latter (i.e. the most complicated case).

Apart from the above WTA problem characteristics, the literature on WTA problem can be categorized into two general groups (see **Table 3**).

1. The first category of approaches includes defensive allocation models where defensive weapons are allocated to targets without taking the behavior of the opposing side into account. The actions of the opposing side are presumed in the scenarios as a given input.

2. The second class of approaches takes the defensive allocation models into account as well as the opposing side's moves. These models often employ the

WTA problem Modeling variation	Number of objective		BF Priority		Engagement policy		Defender positioning	
	Multi-objective	Single-objective	Asset-based	Target-based	Multi-threat	Threat-by-threat	Force coordination	Single platform
SWTA								
DWTA								

Table 2.
The WTA modeling approaches.

Literature category	Reference
Defense allocation models	Newman et al. [16], Benaskeur et al. [18], Zhang et al. [19], Turan [20], Yuming et al. [21], Leboucher et al. [17], Tokgöz and Bulkan [22], Ahner and Parson [4], Kalyanam et al. [8], Davis et al. [23], Naseem et al. [24], Gülpınar et al. [25], Hocaoğlu [26], Kline et al. [27].
Game Models	Matheson [28], Soland [29], Haaland and Wigner [30], Bracken and Brooks [31], Bracken et al. [32], Golany et al. [33], Leboucher et al. [17], Parson [34], Zhang and Zhuang [35].

Table 3.
Classification of the WTA approaches.

two-person-zero-sum-game concept from game theory in the solution process. But we believe that in reality, war is not a zero-sum game and both sides have losses. They reach the

3. solution value of the game by assuming best defensive and offensive moves. The defense wants to minimize the maximum offense's return while the offense acts to maximize the minimum expected return. This approach is more suitable when the inventories of a defender as well as the opposing sides are known to some degree.

This Chapter is focused on defensive scenarios. It is interesting to note that early foundation literature in the WTA field deal mainly on the offensive scenarios, while more recent studies have been focused on defense.

Air defense is a vast area and defensive resource allocation or WTA is a component for which numerous scenarios, models and algorithmic approaches (i.e. techniques for computing solutions) have been proposed. However, an exhaustive categorization for WTA models can be quite large. This section will focus on those aspects that provide for distinctive characteristics of the proposed models. The important characteristics for the modeling and the solution process in the air defense applications are:

1. Simultaneous or sequential attack,

2. Point or area defense systems,

3. Number of attacking and defending weapon types,

4. One or multiple layers of defense,

5. Interceptor WTA policy,

6. Online or Offline application.

In order to better understand the role and importance of these aspects and how to provide a distinctive feature of WTA problem in an air defense system, we shall consider each in further details.

Simultaneous or Sequential Attack: A simultaneous attack is one for which the defense sees all the incoming threats to intercept. The term "known attack size" is also used synonymously for simultaneous attack [36] and is often used in air defense WTA systems on Above Water Warfare (AWW). A sequential attack is the case for which the defense does not know the number of attacking groups in a raid and the number of attackers in each attack group is often used in a ground based air defense context. A mixture of simultaneous and sequential attack may exist for real world situations.

Weapon Types: In a defense scenario in general, attackers and/or defenders may be identical or multiple types. In air defense scenarios in particular, the same pattern holds as well. In practice defender and attackers usually have different types and are equipped with different weapon systems. These attributes usually complicate the WTA problem.

Layered Defense: With reference to the significance of the defended assets, various weapon types are deployed for their protection (i.e. hard-kill and soft-kill weapons). Defense systems of different types, protecting the same targets might have different effective ranges and different functionality. This issue constitutes the layers of defense. In real-word situations, the numbers of these layers are varied and typically have overlaps.

Interceptor Allocation Policy: The defense's interceptor allocation policy such as salvo, shoot-look-shoot and shoot-shoot-look affects the performance and modeling of the air defense system. Defensive systems may have single or multiple engagement opportunities depending on the time–space conditions of the interception and weapon characteristics. Moreover, in WTA literature the defender's interceptors are generally organized into two categories: hard-kill and soft-kill weapons. The hard-kill weapons aim at physically destroying an incoming threat by collision or explosion. The soft-kill weapons aim at diverting the threat from its target using various electronic counter measures [37]. A high volume of research pertains to consideration of hard-kill weapons; generally, this two weapon types are supervised by different control operators. The optimal combining and scheduling of these weapon types, is the responsibility of the higher level commander. This process often is a complex task and usually is studied in the planning systems designing field [37].

Online or Offline Applications: Optimal defensive resource allocation has two main classes of applications, these are, 1) the Online scenario, that is, deploying defensive resources in real time, during real engagements and 2) the Offline scenario, when using allocation algorithms to simulate and model the effectiveness of defensive resources against a given threat scenario. The real-time aspect is one of the major characteristics of real-world air defense situations. Operations research community is the flagship of this area. But despite these initiatives, in many studies in this field real-time aspect of the problem could not be traced.

Reference Features	Point/ Area Defense systems	Weapon Types	One/ Multiple Layers of defense	Interceptor weapon allocation policy	Online/ offline application	Simultaneous/ Sequential attack
Bin Xin et al. [39]	AD	Diff	Mul	S-L-S	Offline	**Seq**
Johansson [5]	AD	Diff	One	S-L	Online	**Simul**
Bin Xin et al. [40]	AD	Diff	Mul	S-L-S	Offline	**Seq**
Zhang et al. [41]	PD	Diff	One	S-L	Online	**Simul**
Turan [20]	PD	Diff	One	S-L	Online	**Simul**
Yuming et al. [21]	PD	Diff	One	S-L	Offline	**Simul**
Leboucher et al. [17]	PD	Diff	Mul	S-L-S	Offline	**Seq**
Tokgöz and Bulkan [22]	AD	Diff	Mul	S-L-S	Offline	**Seq**
Parson [34]	AD	Diff	Mul	S-L-S	Offline	**Seq**
Kalyanam et al. [8]	PD	Diff	Mul	S-L-S	Offline	**Seq**
Naseem et al. [24]	AD	Diff	Mul	S-L-S	Online	**Seq**
Zhang and Zhuang [35]	AD	Diff	Mul	S-L-S	Offline	**Seq**
Kline et al. [27]	AD	Diff	Mul	S-L-S	Offline	**Seq**

Area Defense (AD), Point Defense (PD), Different Weapon Types (Diff), Identical Weapon Types (Iden), Multiple Layers of defense (Mul),One Layers of defense (One), Shot (S), Look (L), Simultaneous attack (Simul), Sequential attack (Seq).

Table 4.
The main characteristics of the WTA approaches.

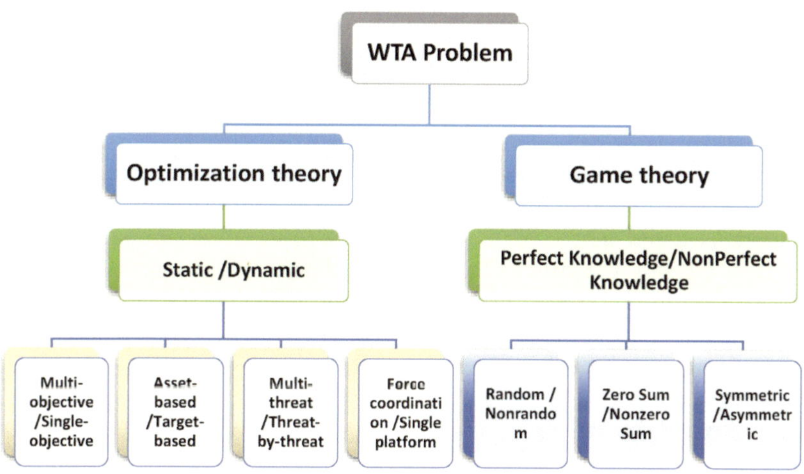

Figure 1.
A general categorization for WTA problem formulation approaches.

The importance of the online scenario is immediate; however, the offline scenario also has significant value and can perceivably be used to aid acquisition or to estimate a measure of preparedness. Further, a capability to consider offline scenarios will most likely enhance the development of online algorithms. This claim follows naturally from the inherent complexity in defensive resource allocation problems, which often necessitates unavoidable approximation for online applications [38].

The classification of related literature that was discussed in preceding section is summarized in **Table 4**. This table contains only the related research which has the above-mentioned properties and those papers which do not talk on their model and their characteristics are excluded. A comprehensive categorization for WTA problem formulation approaches is illustrated in **Figure 1**.

4. Constraints

The WTA problem is subject to various limitations that must be satisfied prior to optimization process. In this section the set of constraints discussed in the WTA literature are reviewed. It is interesting to note that, duo to emergence of time parameter in dynamic mode (i.e. DWTA) the constraints are more diverse.

4.1 SWTA constraints

The following constraints are often raised in the SWTA model and here are called as general constraints:

$$\sum_{i=1}^{|T|} X_{ik} = 1, \quad \forall k \in \{1, 2, \ldots, |W|\} \tag{4}$$

$$\sum_{i=1}^{|T|} X_{ik} \leq 1, \quad \forall k \in \{1, 2, \ldots, |W|\} \tag{5}$$

$$X_{ik} \in \{0, 1\}, \quad \forall i \in \{1, 2, \ldots, |T|\}, \quad \forall k \in \{1, 2, \ldots, |W|\} \tag{6}$$

Eq. (4) means that each firing unit should be assigned to a target and Eq. (6) shows that fractional assignments of firing units to targets are not possible [5]. If the allocation of a certain firing unit to a target is not mandatory the Eq. (4) will be replaced as (5).

4.2 DWTA constraints

The DWTA often divides the total duration of a defensive operation into several discrete time steps in which information is obtained about the allocation outcomes of the previous stages. The DWTA problem aims to assign weapons optimally over time and this feature increases the diversity of constraints in dynamic mode. Therefore, in addition to the limitations listed for SWTA, the following constraints are also considered: capability constraints, strategy constraints, resource constraints, and engagement feasibility constraints.

- Capability constraint:

$$\sum_{i=1}^{|T|} x_{ik}(t) \leq n_k, \quad \forall t \in \{1, 2, \ldots, S\}, \quad \forall k \in \{1, 2, \ldots, |W|\} \tag{7}$$

The constraint set (7) reflects the capability of weapons for firing at multiple targets at the same time and known as capability constraints. Most of actual weapons can shoot only one target at a time. Besides, a special weapon that is capable of engaging multiple targets simultaneously can be viewed as multiple separate weapons. Therefore in most of the WTA literature $n_k = 1$ for $\forall k \in \{1, 2, \dots, W\}$.

- Strategy constraint:

$$\sum_{k=1}^{|W|} x_{ik}(t) \le m_i, \quad \forall t \in \{1, 2, \dots, S\}, \quad \forall i \in \{1, 2, \dots, |\mathrm{T}|\} \tag{8}$$

The constraint set (8) limits the weapon cost for each target at each stage, therefore, they express as strategy constraints. The setting of $m_i \ (i = 1, 2, \dots, \mathrm{T})$ usually depends on the combat performance of available weapons and defender's strategy. For missile-based defense systems and the SLS engagement policy, often $m_i = 1$. For artillery-based defense systems, the value of m_i might be increased greatly under the same demand on defense strength [40].

- Resource constraint:

$$\sum_{t=1}^{S} \sum_{i=1}^{|T|} x_{ik}(t) \le N_k, \quad \forall k \in \{1, 2, \dots, |W|\} \tag{9}$$

The constraint set (9) depicts the amount of ammunition available for weapons and the constraints are known as resource constraints. In DWTA modeling often it is assumed that, the deployment of additional ammunition is not possible in a particular operation.

- Engagement feasibility constraint:

$$\begin{aligned} x_{ik}(t) \le f_{ik}(\mathrm{t}), \quad & \forall t \in \{1, 2, \dots, S\} \\ & \forall k \in \{1, 2, \dots, |W|\}, \quad \forall i \in \{1, 2, \dots, |\mathrm{T}|\} \end{aligned} \tag{10}$$

Constraint set (10) is very important to actual dynamic WTA problems since it takes into account the influence of time windows on the engagement feasibility of weapons (In fact, it represents the difference between static and dynamic modeling). $f_{ik}(\mathrm{t})$ is the indication of actual engagement feasibility for weapon k assigned to target i in stage t. If weapon k can shoot at target i in stage t with any potential reason then $f_{ik}(\mathrm{t}) = 1$; $f_{ik}(\mathrm{t}) = 0$ otherwise [5]. This set of constrains increases the complexity of DWTA problems and the difficulty of generating feasible solutions [5].

In addition to the above constraints, there are a class of technical constrains as well. Constraints such as blind zones (i.e. for fining units and hostile targets), bounds on the survival value of an asset/hostile target [42], scheduling constraints (e.g. constraints such that a weapon has to be shot at a target by a specified time or it is rendered unusable), as well as correlations and heterogeneity between different defensive weapon types (e.g. hard-kill and soft-kill) and so on [43].

5. Algorithmic approach/solution methods

As it is evident from the available literature, a lot of different algorithmic approaches have been suggested for solving the WTA problem in air defense systems. A review of the algorithmic approaches proposed in the literature of WTA is presented in this section and a summary of them is depicted in **Table 5**.

The results show that the WTA problem solving algorithms like many other problems can be divided into three main categories: enumerative techniques, heuristic/approximate techniques and meta-heuristic algorithms. Enumerative techniques are typically guaranteed to find an optimal solution in bounded time for every finite size problem instance; hence, they are referred to as exact algorithms or exhaustive search. The problem with exact algorithms is that the computation times needed often are too high for practical applications. For this reason, in solving the real-world problems with large sizes heuristic methods are commonly used. With heuristic/meta-heuristic algorithms (e.g. guided random search techniques/Genetic Algorithm (GA)), finding an optimal solution is not guaranteed, but instead we can find good solutions in a significantly reduced amount of time [2]. Hence, heuristic/meta-heuristic algorithms are used when we seek good feasible solutions for optimization problems in circumstances where the complexity of the problem or the limited available time for its solution does not allow using the exact algorithms [3]. A summary of the algorithms proposed for WTA is shown in **Table 5**; as mentioned they include exact algorithm, heuristic algorithm as well as meta-heuristic algorithms.

In addition to the algorithmic approaches mentioned in **Table 5**, various variants and combinations of these algorithms also have been suggested in literature for WTA. For example, Khosla [61] developed an algorithm by merging and modifying Simulated Annealing (SA) and GA, Metler and Preston [62] used the solution returned from the MMR algorithm and applied local search on this solution. A GA combined with local search is suggested by Lee et al. [63]. In their study they used Partially Mapped Crossover (PMX), inversion mutation and simulated annealing as local search. In another study they try to resolve the WTA problem with an improved GA, they used greedy reformation scheme so as to have locally optimal offspring (greedy eugenics) which is a kind of novel crossover operator (EX) that try to inherit the good genes from the parents [64]. In the work done by Lu et al. dynamic probability of mutation and crossover in GA have been used to generate new offspring [65]. Shang et al. added an extra step to the GA, after creating new chromosomes by crossover and mutation operators. A local search is applied to create new chromosomes and then evaluates them [66]. They also used a crossover operator called an elite preserving crossover operator to enrich a more effective search. Dou et al. uses chromosomes in matrix representation and adopts the dynamically adjusting punishment gene and self-regulating punishment rates [67]. Li et al. uses matrix-type encoding for

Solving technique	References
Exact algorithms	Malcolm [38], Ahuja et al. [7], Sikanen [44], Karasakal [10], Johansson [5], Turan [20], Bogdanowicz [45], Ahner and Parson [4], Kline [6]
Heuristic algorithms	Newman et al. [16], Tokgöz and Bulkan [22], Yuming et al. [21], Parson [34], Kalyanam et al. [8], Davis et al. [23], Kline [6], Kline et al. [46, 47], Gülpınar et al. [25], Zhang and Zhuang [35], Hocaoğlu [26].
Meta-Heuristic algorithms	Turan [20], Tokgöz and Bulkan [22], Sahin and Leblebicioglu [48], Bogdanowicz et al. [45], Yuming et al. [21], Han et al. [49, 50], Ahner and Parson [4], Pendharkar [51], Davis et al. [23], Gong et al. [52], Klinkowski et al. [53], Gülpınar et al. [25], Wang et al. [54], Ma et al. [55], Kline et al. [27], Rudek and Heppner [43], Kalaiselvi and Selvi [56], Shooli and Javidi [57], Zhu et al. [58], Xu et al. [59], Ejaz et al. [60]

Table 5.
Classification of the algorithmic approaches used to solve the WTA models.

chromosomes and a new crossover/mutation operator called "circle swap" [68]. Sahin and Leblebicioglu applied fuzzy reasoning to approximate optimum allocations in real-time for use on a battlefield [48]. Lagrangian relaxation was proposed by Yu Z et al. to decompose the problem into two tractable sub-problems while iteratively updating the Lagrange multipliers [69].

Similar to the SWTA, numerous methods have been employed to provide solutions for various types of DWTA problems. The DWTA divides the total duration of a defensive operation into several discrete time steps in which information is obtained about the allocation outcomes of the previous stages. Any hostile targets destroyed during a stage will be no longer targeted in subsequent stages, allowing the operator to make better use of their weapons. Hosein is an originator of the dynamic instance. He provided several results which are generalizable to the DWTA problem [13]. Murphey used stochastic decomposition for the two-stage DWTA problem [15]. Chang used a static WTA approximation scheme within an iterative linear network flow framework to efficiently provide high-quality solutions for the DWTA [70]. Wu et al. applied a modified GA to the DWTA and introduced weapon use deadlines within the problem formulation [71]. Xin et al. developed a heuristic which uses problem information (domain knowledge) and constraint programming to assign priorities to assignments [40]. Evolutionary heuristics which use a hybridized GA with memetic algorithms have also been applied by Chen et al. [72]. Other researches have been done to solve the WTA problem which are listed in **Table 5**.

Though it has not been researched to the extent of the SWTA problem, the DWTA problem provides a more practical implementation by including a temporal component. As such, the DWTA is a much more complex problem from a mathematical standpoint and has received a fair amount of attention in the literature. Due to the increasing computing power of computers and progress of science such as artificial intelligence (AI), researchers are moved towards the use of meta-heuristic algorithms to solving the WTA especially for DWTA problem.

6. Conclusions

In this Chapter, the literature in the area of WTA system with the emphasis on the modeling and solving approaches has been reviewed. Based on modeling approaches the literature has been classified into two groups. The first class of approaches allocates the defensive sources to targets without taking the behavior of the opposing side into account. The second class of approaches considers the opposing side's moves as well as the defensive moves. In the second part, the important characteristics for the WTA models and the solution process for solving the WTA problem are reviewed. The results show that solving algorithms, like many other problems, can be divided in three main categories: enumerative techniques (i.e. exact or exhaustive methods), heuristic/approximate techniques and the meta-heuristic ones.

Author details

Abdolreza Asadi Ghanbari[1]*, Mousa Mohammadnia[2], S. Abbas Sadatinejad[3] and Hossein Alaei[4]

1 Department of Computer Engineering, Science and Research Branch, Islamic Azad University, Tehran, Iran

2 Academy of Artificial Intelligence and New Technologies, Tehran, Iran

3 School of Computer Engineering, Iran University of Science and Technology, Tehran, Iran

4 Faculty of Entrepreneurship, Tehran University, Tehran, Iran

*Address all correspondence to: asadi.abdolreza@gmail.com

References

[1] Manne AS. A Target-Assignment Problem. Operations Research. 1958; 6 (3): 346–351.

[2] Ghanbari, A. A; Alaei, H. Meta-Heuristic Algorithms for Resource Management in Crisis Based on OWA Approach. Appl Intell. 2020; https://doi.org/10.1007/s10489-020-01808-y.

[3] Ghanbari, A. A.; Alaei, H.; Mohammadnia, M. A Multi-Stage Modelling Approche for Allocation of Defense Resources to Invading Targets. Adv. Defence Sci.Technol. 2020; 2:167–173. (In Persian)

[4] Ahner DK, Parson CR. Optimal multi-stage allocation of weapons to targets using adaptive dynamic programming. Optimization Letters. 2015; 9(8):1689–1701.

[5] Johansson F. Evaluating the Performance of TEWA Systems [thesis]. Skövde: University of Skövde; 2010.

[6] Kline A. Real-time heuristic algorithms for the static weapon-target assignment problem. Master's thesis, Air Force Institute of Technology; 2017.

[7] Ahuja RK, Kumar A, Jha KC, Orlin JB. Exact and Heuristic Methods for the Weapon Target Assignment Problem. Operations Research. 2007; 55 (6):1136–1146.

[8] Kalyanam K, Rathinam S, Casbeer D, Pachter M. Optimal Threshold Policy for Sequential Weapon Target Assignment. IFAC-PapersOnLine. 2016; 49(17):7–10.

[9] Fujita H, Gaeta A, Loia V, Orciuoli F. Hypotheses Analysis and Assessment in counter-terrorism activities: a method based on OWA and Fuzzy Probabilistic Rough Sets. IEEE Transactions on Fuzzy Systems; 2019. DOI: 10.1109/TFUZZ.2019.2955047.

[10] Karasakal O. Air defense missile-target allocation models for a naval task group. Computers & Operations Research. 2008; 35(6):1759–1770.

[11] DenBroader GG, Ellison RE, Emerling L. On optimum target assignments. Operations Research. 1958; 7(3):322–326.

[12] Orlin D. Optimal Weapon Allocation Against Layered Defenses. Naval Research Logistics. 1987; 34(5): 605–617.

[13] Hosein PA. A class of dynamic nonlinear resource allocation problems. Tech. Rep. LIDS-TH-1922, Massachusetts Inst Of Tech Cambridge Lab For Information And Decision Systems; 1989.

[14] Huaiping C, Jingxu L, Yingwu C, Hao W. Survey of the research on dynamic weapon-target assignment problem. Journal of Systems Engineering and Electronics. 2006; 17 (3):559–565.

[15] Murphey RA. An approximate algorithm for a weapon target assignment stochastic program. In: Approximation and Complexity in Numerical Optimization. Springer. 2000; 406–421.

[16] Newman AM, Rosenthal RE, Salmeron J, Brown GG, Price W, Rowe A, Fennemore CF, Taft RL. Optimizing Assignment of Tomahawk Cruise Missile Missions to Firing Units. Naval Research Logistics. 2011; 58(3): 281–295.

[17] C Leboucher, H-S Shin, S Le Menec, A Tsourdos, A Kotenkoff ,P Siarry, R Chelouah. Novel Evolutionary Game Based Multi-Objective Optimisation for Dynamic Weapon Target Assignment. Preprints of the 19th World Congress, The International Federation of

Automatic Control, Cape Town, South Africa, 2014.

[18] Benaskeur AR, Irandoust H, Kabanza F, Beaudry E. Decision Support Tool for Anti-Ship Missile Defence Operations. Paper presented at: ICCRTS 2011. Proceedings of the 16th International Command and Control Research and Technology Symposium; June 21–23; Quebec City, Quebec, Canada; 2011.

[19] Zhang J, Xu C, Wang X, Yuan D. ACGA Algorithm of Solving Weapon-Target Assignment Problem. Open Journal of Applied Sciences. 2012; 74–77.

[20] Turan A. Algorithms for the weapon-target allocation problem. Master's thesis, The Middle East technical university, Turkey; 2012.

[21] Yuming LU, Miao W, Ming LI. The Air Defense missile Optimum Target Assignment Based on the Improved Genetic Algorithm. Journal of Theoretical and Applied Information Technology. 2013; 48(2):809–816.

[22] Tokgöz A, Bulkan S. Weapon Target Assignment with Combinatorial Optimization Techniques. (IJARAI) International Journal of Advanced Research in Artificial Intelligence. 2013; 2(7):39–50.

[23] Davis MT, Robbins MJ, Lunday BJ. Approximate dynamic programming for missile defense interceptor fire control. European Journal of Operational Research. 2017; 259(3):873–886.

[24] Naseem A, Khan SA, Malik AW. A real-time man-in-loop threat evaluation and resource assignment in defense. Journal of the Operational Research Society. 2017; 68(6):725–738.

[25] Gülpınar N, Çanakoğlu E, Branke J. Heuristics for the stochastic dynamic task-resource allocation problem with retry opportunities. European Journal of Operational Research. 2018; 266(1):291–303.

[26] Hocaoğlu MF. Weapon target assignment optimization for land based multi-air defense systems: A goal programming approach. Computers & Industrial Engineering. 2019; 128:681–689.

[27] Kline AG, Ahner DK, Hill R. The Weapon-Target Assignment Problem. Computers & Operations Research. 2019; 105:226–236.

[28] Matheson JD. Preferential Strategies with Imperfect Weapons. AR 67–1, Analytic Services Inc, Arlington, VA; 1967.

[29] Soland RM. Optimal Defensive Missile Allocation: A Discrete Min-Max Problem. Operations Research. 1973; 21 (2):590–596.

[30] Haaland CM, Wigner EP. Defense of Cities by Antiballistic Missiles. SIAM Review. 1977; 19(2):279–296.

[31] Bracken J, Brooks PS. Attack and Defense of ICBMs Deceptively Based in a Number of Identical Areas. Naval Research Logistics Quarterly. 1985; 32 (2):193–207.

[32] Bracken J, Falk JE, Allen TAJ. Robustness of Preallocated Preferential Defense With Assumed Attack Size and Perfect Attacking and Defending Weapons. Naval Research Logistics. 1987; 34(1):23–41.

[33] Golany B, Goldberg N, Rothblum UG. Allocating multiple defensive resources in a zero-sum game setting. Annals of Operations Research. 2012; 225(1):91–109.

[34] Parson CR. Approximate Dynamic Programming For Military Resource Allocation. PhD Thesis, Air Force Institute of Technology, Ohio; 2014.

[35] Zhang J, Zhuang J. Modeling a multi-target attacker-defender game

with multiple attack types. Reliability Engineering & System Safety. 2019; 185: 465–475.

[36] Karasakal O. Optimal Air Defense Strategies for a Naval Task Group. phd thesis, The Middle East technical universit, Turkey; 2004.

[37] Anissa Frini, Adel Guitouni, Abder Rezak Benaskeur, Éloi Bossé. Single ship resource allocation in above water warfare. Technical Report TR 2006–766, DRDC Valcartier, 2008.

[38] Malcolm WP. On The Character and Complexity of Certain Defensive Resource Allocation Problems. Weapons Systems Division Systems Sciences Laboratory, Edinburgh South Australia; 2004.

[39] Xin B, Chen J, Zhang J, Dou L, Peng Z. Efficient Decision Makings for Dynamic Weapon-Target Assignment by Virtual Permutation and Tabu Search Heuristics. IEEE Transactions on Systems. Man, and Cybernetics, Part C (Applications and Reviews). 2010; 40 (6):649–662.

[40] Xin B, Chen J, Peng Z, Dou L, Zhang J. An Efficient Rule-Based Constructive Heuristic to Solve Dynamic Weapon-Target Assignment Problem. IEEE Transactions on Systems, Man, and Cybernetics - Part A: Systems and Humans. 2011; 41(3):598–606.

[41] Zhang J, Xu C, Wang X, Yuan D. ACGA Algorithm of Solving Weapon-Target Assignment Problem. Open Journal of Applied Sciences. 2012; 74–77.

[42] Taner G. Weapon-Target Allocation and Scheduling Air Defense with Time Varying Hit Probabilities. Master's thesis, The Middle East technical university, Turkey; 2007.

[43] Rudek R, Heppner L. Efficient algorithms for discrete resource allocation problems under degressively proportional constraints. Expert Systems with Applications. 2020; https://doi.org/10.1016/j.eswa.2020.113293.

[44] Sikanen T. Solving weapon target assignment problem with dynamic programming. Technical report, Tech. Rep. 55670; 2008.

[45] Bogdanowicz ZR, Tolano A, Patel K, Coleman NP. Optimization of weapon–target pairings based on kill probabilities. IEEE transactions on cybernetics. 2013; 43(6):1835–1844.

[46] Kline A, Ahner D, Lunday B. A heuristic and metaheuristic approach to the static weapon target assignment problem. Tech. Rep. COA-01-17, Air Force Institute of Technology Center for Operational Analysis; 2017a.

[47] Kline A, Ahner D, Pachter M. A greedy hungarian algorithm for the weapon-target assignment problem. Tech. Rep. COA-02-17, Air Force Institute of Technology Center for Operational Analysis; 2017b.

[48] Sahin MA, Leblebicioğlub K. Approximating the optimal mapping for weapon target assignment by fuzzy reasoning. Journal of Information Sciences. 2013; 255:30–34.

[49] Han Y, Gong D, Sun X. A discrete artificial bee colony algorithm incorporating differential evolution for the flow-shop scheduling problem with blocking. Engineering Optimization. 2015; 47(7):927–946.

[50] Han Y, Gong D, Sun XY, Pan Q. An improved NSGA-II algorithm for multi-objective lot-streaming flow shop scheduling problem. International Journal of Production Research. 2014; 52 (8):2211–2231.

[51] Pendharkar C. An ant colony optimization heuristic for constrained task allocation problem. Journal of Computational Science. 2015; 7:37–47.

[52] Gong D, Han Y, Sun J. A novel hybrid multi-objective artificial bee colony algorithm for blocking lot-streaming flow shop scheduling problems. Knowledge-Based Systems. 2018; 148:115–130.

[53] Klinkowski M, Lechowicz P, Walkowiak K. Survey of resource allocation schemes and algorithms in spectrally-spatially flexible optical networking. Optical Switching and Networking. 2018; 27:58–78.

[54] Wang J, Hu X, Demeulemeester E, Zhao Y. A bi-objective robust resource allocation model for the RCPSP considering resource transfer costs. International Journal of Production Research. 2019; https://doi.org/10.1080/00207543.2019.1695168.

[55] Ma K, Liu X, Li G, Hu S, Guan X. Resource allocation for smart grid communication based on a multi-swarm artificial bee colony algorithm with cooperative learning. Engineering Applications of Artificial Intelligence. 2019; 81:29–36.

[56] Kalaiselvi S, Selvi CSK. Hybrid Cloud Resource Provisioning (HCRP) Algorithm for Optimal Resource Allocation Using MKFCM and Bat Algorithm. Wireless Personal Communications. 2020; 111:1171–1185.

[57] Shooli RG, Javidi MM. Using gravitational search algorithm enhanced by fuzzy for resource allocation in cloud computing environments. SN Applied Sciences. 2020; https://doi.org/10.1007/s42452-020-2014-y.

[58] Zhu Z, Peng J, Liu K, Zhang X. A game-based resource pricing and allocation mechanism for profit maximization in cloud computing. Soft Computing. 2020; 24:4191–4203.

[59] Xu X, Hao J, Zheng Y. Multi-objective artificial bee colony algorithm for multi-stage resource leveling problem in sharing logistics network. Computers & Industrial Engineering. 2020; https://doi.org/10.1016/j.cie.2020.106338.

[60] Ejaz W, Sharma SK, Saadat S, Naeem M, Chughtai NA. A comprehensive survey on resource allocation for CRAN in 5G and beyond networks. Journal of Network and Computer Applications. 2020; https://doi.org/10.1016/j.jnca.2020.102638.

[61] Khosla D. Hybrid genetic approach for the dynamic weapon-target allocation problem, Proc. SPIE. 2001; 4396:244–259.

[62] Metler WA, Preston FL. A suite of weapon assignment algorithms for a SDI mid-course battle manager. Technical report.Naval Research Laboratory. Washington, DC; 1990.

[63] Lee ZJ, Lee CY, Su SF. An immunity-based ant colony optimization algorithm for solving weapon–target assignment problem. Applied Soft Computing. 2002a; 2(1):39–47.

[64] Lee ZJ, Su SF, Lee CY. Efficiently solving general weapon-target assignment problem by genetic algorithms with greedy eugenics. IEEE Transactions on Systems, Man, and Cybernetics, Part B (Cybernetics). 2003; 33(1):113–121.

[65] Lu H, Zhang H, Zhang X, Han R. An improved genetic algorithm for target assignment optimization of naval fleet air defense. in Proc. 6th World Congr. Intell. Control Autom., Dalian, China. 2006, p. 3401–3405.

[66] Shang G, Zaiyue Z, Xiaoru Z, Cungen C. Immune Genetic Algorithm for Weapon-Target Assignment Problem. Workshop on Intelligent Information Technology Application, Zhang Jiajie; 2007. pp. 145–148.

[67] Dou Jihua, Yang Xingbao and Lu Yonghong. Improved Genetic

Algorithm For Multichannel Ship-To-Air Missile Weapon System WTA Problem. 2nd IEEE International Conference on Computer Science and Information Technology, Beijing, 2009, pp. 210–214

[68] Li P, Wu L, Lu F. A mutation-based GA for weapon-target allocation problem subject to spatial constraints. Paper presented at: ISA 2009. Proceedings of the Intelligent Systems and Applications; May 23–24, Wuhan, China; 2009.

[69] Ni M, Yu Z, Ma F, Wu X. A Lagrange relaxation method for solving weapon-target assignment problem. Mathematical Problems in Engineering. 2011; 1024–1033

[70] Chang SC, James RM, Shaw JJ. Assignment algorithm for kinetic energy weapons in boost defence. In Proceedings of the IEEE 26th Conference on Decision and Control, Los Angeles, California, 1987.

[71] Wu L, Wang HY, Lu FX, Jia P. An anytime algorithm based on modified GA for dynamic weapon-target allocation problem. IEEE Congress on Evolutionary Computation (IEEE World Congress on Computational Intelligence), Hong Kong; 2008 pp. 2020–2025.

[72] Chen j, Bin Xin, ZhiHong Peng, LiHua Dou, Juan Zhang. Evolutionary decision-makings for the dynamic weapon-target assignment problem. Science in China Series F: Information Sciences. 2009; 52(11): 2006–2018.

Chapter 6

A Review on Advanced Manufacturing Techniques and Their Applications

Anand J. Patel and Gülenay A. Kilic

Abstract

Advancement in manufacturing processes has drawn preeminent interest from researchers and industry, it makes the process of manufacturing more productive and capable of high efficiency. Advancement of technology has been done by several approaches to combine different manufacturing processes with similar objectives of increasing material removal rate, improving surface integrity, reducing tool wear, reducing production time, and extending application areas. A combination of different processes has been called 'Hybrid' processes by various researchers, engineers, and industry expert. Hybrid processes open new opportunities and applications for manufacturing various components that are not able to be produced economically by processes on their own. This review report starts with the classification of current manufacturing processes based on the nature of the processing. The main part of this report is reviews of existing and widely used manufacturing processes that recently reported in a decade. Purpose of this report to produce an overview of these different processes by reviewing various research papers.

Keywords: manufacturing processes, advancement of technology, hybrid processes, economic processes, recent manufacturing process

1. Introduction

In the present manufacturing industry developed a new approach to manufacture products. Generally conventional manufacturing processes which has been widely adopted are computer numerical control (CNC) machining, transformative processes such as forming, joining and dividing operations, for example, welding and sawing. Also, additive manufacturing has been adopted in many sectors of industries.

Conventional manufacturing processes, however, have their inherent drawbacks which cannot be eliminated. In other words, due to their technological constraints, they are not always feasible to produce various components in terms of geometry, dimension, and strength, etc. CNC machining can have difficulties in machining complex shapes due to tool accessibility. High temperature and tool wear are other considerations while machining hard materials. As compared to CNC machining, Rapid prototyping is still restricted because of long production time and low accuracy. Limited materials' formability and spring-back effect confine the development of forming processes. In welding processes, furthermore, dimensional precision is hard to completely control.

In view of the issues referenced above, a combination of two or more manufacturing processes with different manufacturing principle is a new topic for many researchers, such processes are called hybrid manufacturing. The main driven idea to develop such process is to minimize individual manufacturing limitation and enhance their advantages by combining two or more manufacturing processes. The combination of CNC machining and additive processes may provide a new substantial solution to the limitations of additive processes due to the high accuracy and machining speed that machining processes offer. Moreover, the combination of laser heating and forming reduces spring back behavior. The combination of drilling and ultrasonic vibration can reduce the tool wear rate and cutting force. The involvement of laser drilling and electrochemical machining (ECM) significantly removes the recast layer and heat affect zone.

Hybrid manufacturing has enormous potential to produce more complex parts, provide more tolerance of flexibility while maintaining high accuracy in relatively lower production time. This is now becoming trend among researchers to develop hybrid processes for enhancing processes capabilities, minimizing their limitations and broadening application areas.

The major topic contains in this book chapter are classification of existing manufacturing processes and its technologies based on their nature of operation; simplification of different strategies and terms used by researchers; define, identify and classification of hybrid manufacturing processes.

2. Need and significance of advance manufacturing

2.1 Major drivers of advanced manufacturing

As shown in **Figure 1** Advance Manufacturing has three major driver's material driver, process driver and operational driver. A development in new materials which are hard to machines are categorized under material driven processes, while process driver is due to specific product necessities such as precision, high accuracy,

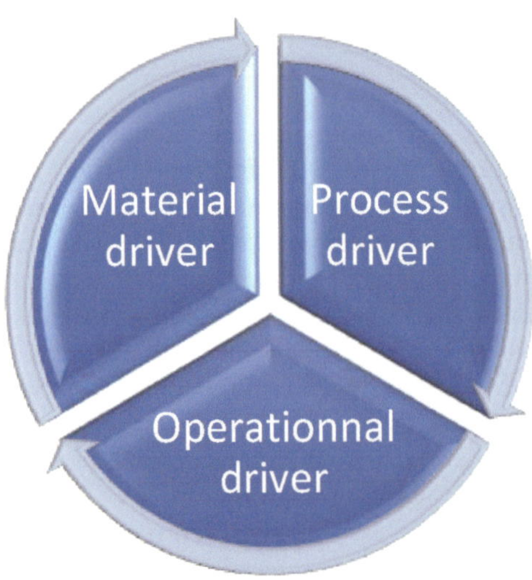

Figure 1.
The major drives of advanced manufacturing processes.

Figure 2.
Manufacturing philosophy wise the trends.

high quality etc. The operation driver is responsible for shrinking time to market requirements which led to high production rate, apart from that its attributed to produce products cost effective by cutting manufacturing costs.

2.2 Manufacturing philosophy wise the trends

Over a last few decade's manufacturing theory wise the trends can be observed in **Figure 2** and those are as follows, In the 1960's a measure trend for manufacturing processes were cost-driven. Then it became quality driven in 1980. However recently cost quality and time to market are the most important drivers in manufacturing industries.

The need for advanced manufacturing can be attributed to the following.

a. Limitations in conventional methods.

b. Rapid improvements in material properties.

c. High tolerance requirements, product requirements.

Some major advantages of advanced manufacturing processes

a. Production with low cost

b. Automated data transmission

c. Miniaturization

d. Precise and ultra-precession Finishing

3. Classification of advanced manufacturing techniques

3.1 General classification of advanced manufacturing techniques

A general classification is described in **Figure 3**. Techniques like ultrasonic machining, water jet machining, abrasive water jet machining, abrasive jet machining, abrasive flow machining, magnetic abrasive flow machining are classified under Mechanical category. Plasma earth machining, laser beam processing, electron beam processing, electric discharge machining, ion beam machining etcetera

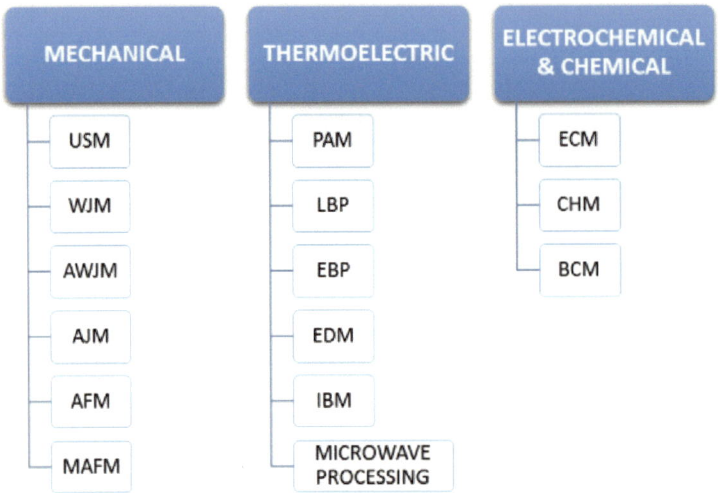

Figure 3.
General classification of advanced manufacturing techniques.

are classified under Thermoelectric category. One new process is emerging in this category, this is microwave processing. Microwave processing of materials have been extensively used, especially in polymeric and ceramic material processing [1].

The new trend in advanced manufacturing is hybridization of two or more than two manufacturing processes. The International Academy for Production Engineering (CIRP) has suggested two definition based on its nature of Hybrid Manufacturing.

i. "A hybrid manufacturing process combines two or more manufacturing processes into a new combined set-up whereby the advantages of each discrete process can be exploited synergistically" (Open definition).

ii. "A hybrid manufacturing process comprises a simultaneous acting of different processing principles on the same zone" (Narrow definition).

3.2 Broad classification of hybrid manufacturing techniques

A classification has been done based on process nature and further its categorized in **Figure 4**. A classification by Kalpakjain and Schmid (2010) is as follows.

i. Joining Technology: Include processes which allowed two or more workpieces to join and to form a new workpiece. Examples are welding and assembly.

ii. Dividing Technology: Contain processes which are the opposite to the joining processes, for instance, sawing and disassembly.

iii. Subtractive Technology: Contains operation which are responsible for material removal, in an order to make new workpiece materials has removed from a single workpiece, all machining operations are the examples of subtractive technology.

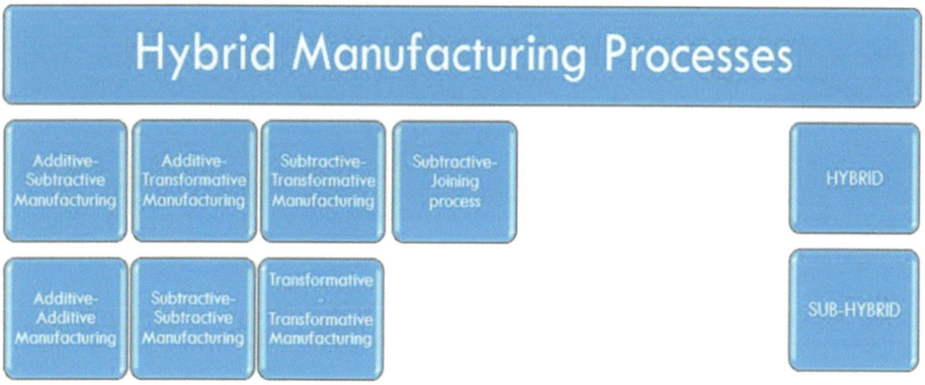

Figure 4.
Classification of hybrid and sub hybrid manufacturing processes.

iv. Transformative Technology: Without changing its mass a workpiece is transformed to create another new workpiece. Examples of transformative processes are forming, heat treatment and cryogenic cooling.

v. Additive Technology: To build a new workpiece material must be added to an existing workpiece which result in an increment in the mass of the final workpiece. The existing technology such as rapid prototyping processes, injection molding and die casting are the example of additive manufacturing processes.

Based on above classification a sub-categorization of hybrid and sub-hybrid processes are mentioned in **Figures 5** and **6** respectively.

3.3 Sub categorization of hybrid manufacturing processes

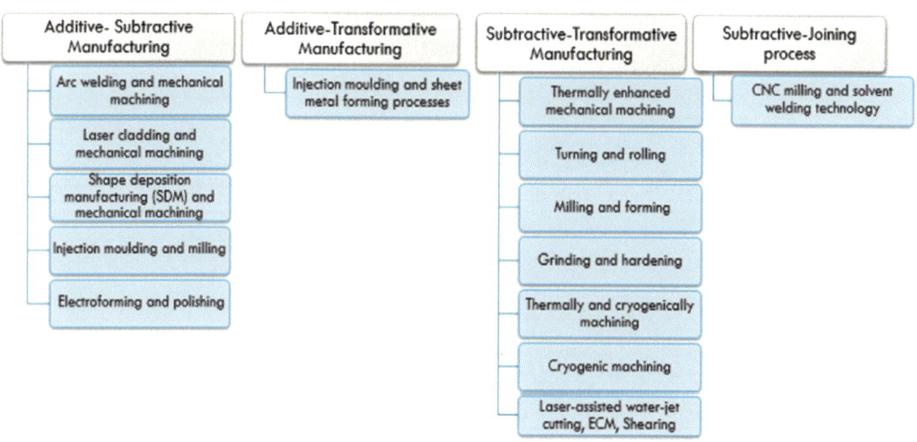

Figure 5.
Sub categorization of hybrid manufacturing processes.

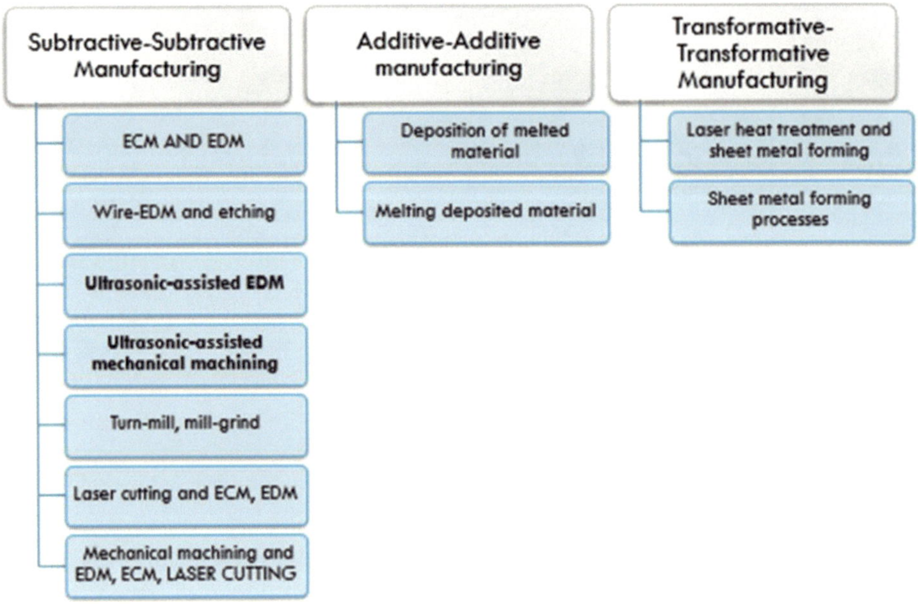

Figure 6.
Sub categorization of sub hybrid manufacturing processes.

4. Overview of processes (setup, procedures, manufacturing)

4.1 Additive subtractive manufacturing processes

4.1.1 Arc welding & mechanical machining

In this process a 3D welding is used as an additive manufacturing process while milling process is used as a material removal process, single beads of welding is deposited side by side by using conventional gas metal arc welding. By controlling the welding parameters mainly speed and power, the thickness of bead can be set in a range between 0.5 and 1.5 mm. After deposition of a bead layer, surface of layer is machined by using milling to achieve a smooth surface with defined thickness for

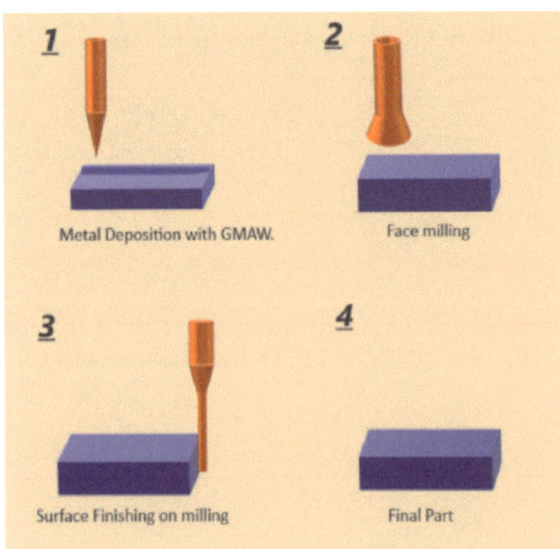

Figure 7.
Gmaw and milling operation [2].

next weld bead deposition, as shown in **Figure 7**. The combination of this process of welding pass and with face milling offers a advantage in controlling layer thicknesses range between A0.1 and A1 mm. After the sequence of weld bead deposition and face milling is finished, surface finishing operation is take place on same machining setup in order to subtract the left stair steps pattern on the surface of the machining part and to improve the accuracy of the near-net shape metal part [2–5].

4.1.2 Laser cladding and mechanical machining

The combination of selective laser cladding (SLC) and milling process result in a hybrid process and its design and construction is shown in **Figure 8**. By laser cladding a layer of material is deposited on the workpiece, and then subsequently milling operation was introduced to smooth the deposited clad surface. After the cladding operation, the engaged focal point of the laser was moved away and the workpiece was moved to the position beneath the milling head. Then, to achieve the desired accuracy and a smooth surface, the top surface of the workpiece with the deposited clad profile was machined in order to make the surface ready for the next cladding operation [6–10].

4.1.3 Shape deposition manufacturing (SDM) and mechanical machining

One of the variations of Shape Deposition Manufacturing is a Mold SDM and various process steps can be seen in **Figure 9**. SDM is a process which include additive and subtractive manufacturing process which is used to manufacture a various metal based and polymer-based parts. Almost every layer deposition technique breaks down the model into moderately thin and uniform thickness layers. However, in the process of shape deposition manufacturing layers are 3D, also it can be of arbitrary thickness and it has not compulsion to be planar. Such a decomposition in additive process allows the quantity of layers to be limited which is leads to reduction in processing time. In Mold based Shape Deposition Method molds are constructed using SDM process, afterwards these are used to cast a various part material. For illustration, the Mold Shape Deposition Methods construct succession for a basic part with three layers. The shape of the mold cavity is defined by the support material segments. The mold itself is formed by the support material which are constructed around by the segments of mold material. In initial step the mold is constructed up layer by layer by Deposition Methods techniques. In later step support material is removed which allowed a mold ready for casting. In subsequent

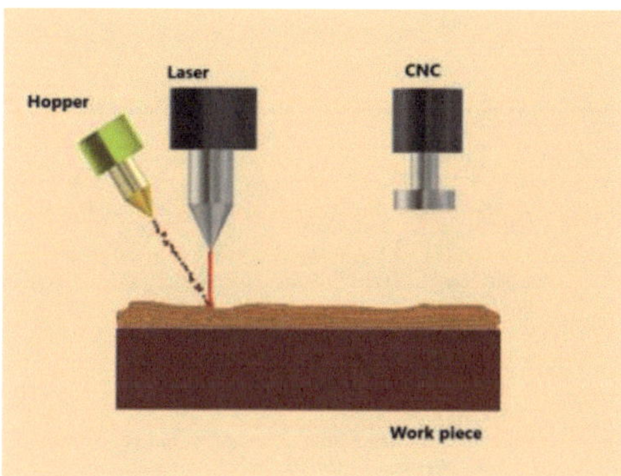

Figure 8.
Laser cladding and mechanical machining [6].

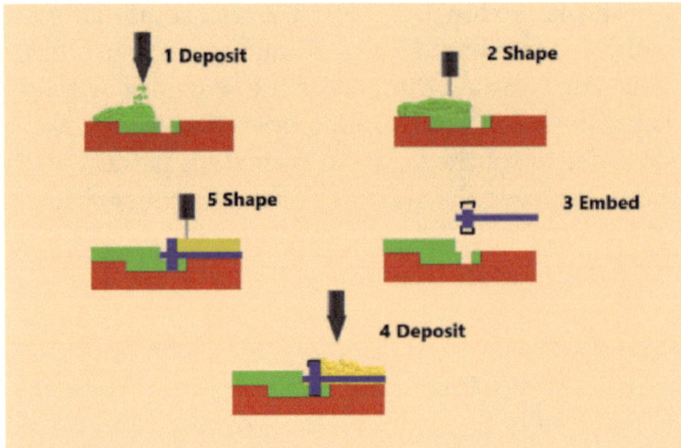

Figure 9.
SDM and machining.

stage after part material cured it removed from the mold and finishing operation take place, for example removal of gates and runners. An alternative method also can be applicable in which the mold material is used as a fixturing during finishing operations and then it can be remove in later stage [11–14].

4.1.4 Injection molding and machining

Klelkar et al. developed a re-configurable molding process by using movable pins, as shown in **Figure 10**, to generate a cavity in the mold, the process has limitation as a part surface is only approximated. According to the change of product design the pins can be re-positioned, and a required mold cavity can be produced. Kelkar and Koc introduced multiaxis machining in re-configurable mold tooling. Multi-axis machining was used to improve the surface accuracy of the part after a part is molded. The both process was carried on same setup and make it suitable for a batch production [15, 16].

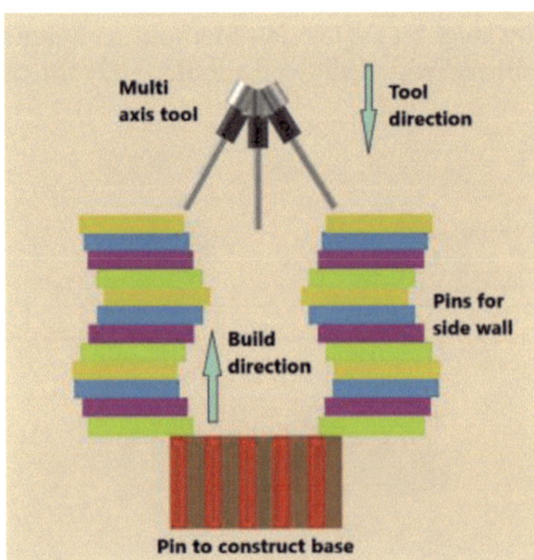

Figure 10.
Injectioun molding and machining.

4.2 Additive and transformative manufacturing processes

4.2.1 Injection molding and sheet metal forming

Polymer Injection Forming (PIF) is a one-operation manufacturing technique and a novel approach to manufacture sheet metal/polymer macro-composite. During the process, the pressure which is exerted by the injection of the molten polymer is used to shape a metal sheet inside an injection mold. In the same step permanent bond between the metal sheet and polymer creating a fully finished product is carried out in only one manufacturing step. The Polymer Injection Forming manufacturing is a combination of the metal forming and injection molding processes. The process sequence can be seen in **Figure 11**.

i. Between the open halves of a mold a metal sheet is inserted.

ii. When the mold closes, the metal sheet is cut down to the defined blank size. During the same operation and on same apparatus the blank can be shaped by bending the sheet or by deep drawing metal into the basic product shape.

iii. The polymer is injected into the remaining cavity when the mold is completely closed and a second, hydrostatic, deformation step is applied to shape the metal sheet into its definite form. In this phase of the PIF process, the physical adhesion between the metal sheet and the injected polymer is obtained.

iv. Finalizing the production cycle, the product is ready and is removed from the mold.

4.3 Subtractive and transformative manufacturing processes

4.3.1 Thermally enhanced mechanical machining

In the process of thermally enhanced mechanical machining externally heat sources is applied to heat the metal locally in front of the cutting tool. The heating effect changes the microstructure of the workpiece and it softened the material which allowed reduction in hardness, cutting forces and tool wear during material removing process on conventional machines. The frequently used external heat sources are laser beam and plasma. Laser assisted mechanical machining has been considered as an alternative process for machining of high-strength materials, such as high-temperature alloys, metal matrix composites, ceramics [18–21].

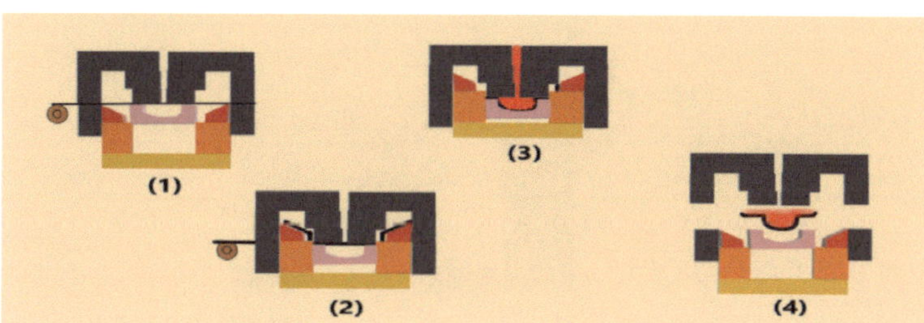

Figure 11.
Polymer Injection Forming process sequence [17].

4.3.2 Grinding and hardening

The grind-hardening process shown in **Figure 12** used to surface harden steel parts. This process is using greater depth of cuts, up to 1 mm, and slower work-piece speeds. Rotating grinding wheel generate a heat at contact by which the surface temperature of material raised above that of authentication. Also by introducing self-quenching for heat dissipation by using coolant, martensitic phase transformation takes place [22, 23].

4.4 Subtractive joining process

A configuration of schematic diagram of Solvent Welding Freeform Fabrication Technique (SWIFT) process is shown in **Figure 13**. In initial stage a standard size solvent weldable thermoplastic sheets placed loaded into a sheet feeder. The most frequently used solvent weldable thermoplastics materials are polystyrene, polycarbonate, PVC, and ABS. A pair of pinch rollers are used to fed sheets forward. Forwared sheet passes through a solvent masking which assist to prevent unwanted welding in desired areas. Solvent has been applied to the underside of the sheet at solvent masking station. Sheets are feeding continuous until it is positioned over the build platform. A platen is used to apply pressure on the sheet which is positioned over the stack of previously assembled sheets. In the process the new sheet placed on the top surface of the previously applied sheet, then the applied solvent between

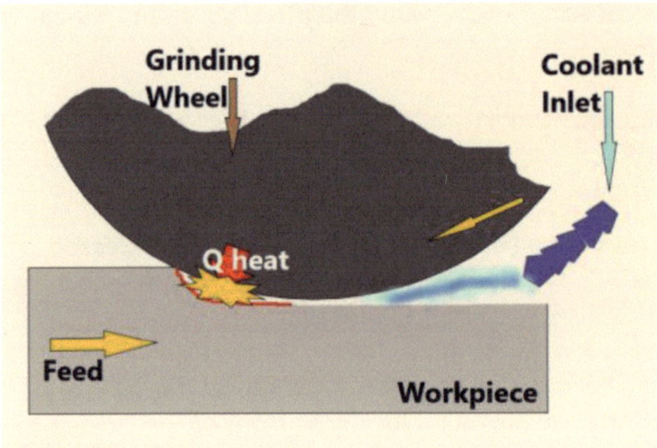

Figure 12.
Grind hardning process.

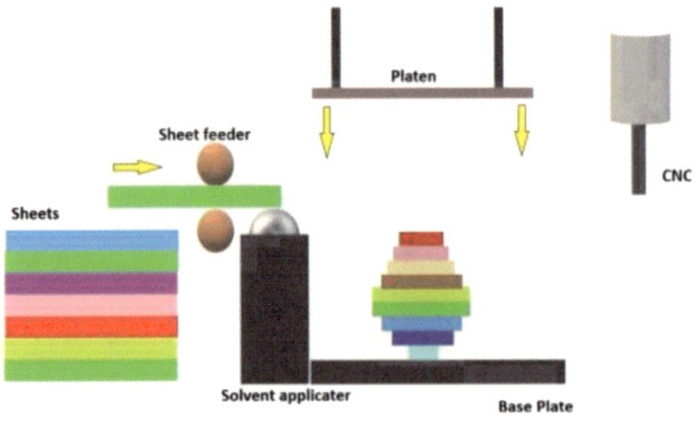

Figure 13.
Swift process configuration.

the two sheets at the interface dissolves polymer chains. After a while as solvent evaporates or absorbed it result in an interface weld and a solid part, which allows new molecular bonds between the two sheets to be established. This process is very rapid, and in practical, the platen apply pressure for only 5 seconds. Then by a CNC milling machine the cross-sectional contour cut down with a small diameter end mill cutter. As the machined sheets are preassembled prior to machining, process does not have any requirement of post assembly or special registration. Final product is completed after every layers has been added to the assembly and machined, the part can be removed from the build platform for the post processing as per the requirement [24, 25].

4.5 Additive manufacturing processes

4.5.1 Melting deposited material

Plasma laser deposition manufacturing working principle is shown in **Figure 14**. Powder is continuously feeding by control system into molten pool where powder is melted due to the focused laser beam and it result in re-solidification. The coatings thickness and 3D CAD model part must be predefined. To reduce oxidation, the whole operation is carried out under inert gas argon environment. The worktable is controlled by the control system. The material is deposited side by side with a defined amount of overlap. After deposition of an entire coating layer, the beam height of laser and plasma considerably keep away from the surface. The variables in the process such as power of combine plasma and laser beam, velocity of beam movement, feeding rate of material are controlled as per requirement of part to manufacture a final product.

This proposed PLDM process offers several advantages over conventional surface coatings techniques [26, 27]:

i. The traditional spray coating process can only used for thin coatings; The proposed PLDM process can allow to deposit thick layer coatings and it surfaces coating thickness can be controlled according to their requirements.

ii. The PLDM also allows material with a higher melting point such as alloys and refractory materials to directly fabricating different materials coatings.

iii. The PLDM process also offers comparatively high-quality microstructure and other mechanical characteristics.

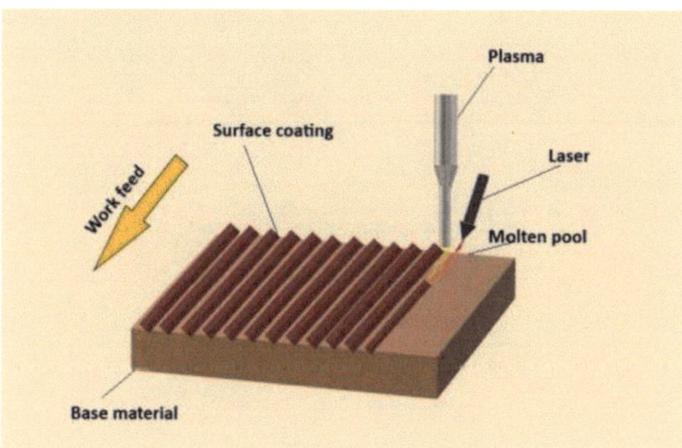

Figure 14.
Plasma laser deposition manufacturing.

iv. The PLDM process is widely adopted metal based rapid prototyping and tooling requirements.

4.5.2 Deposition of melted material

The laser deposition is one of the widely used metal deposition processes. This process offers a several advantages over other conventional solid deposition techniques. The advantages such as robust deposition with more accuracy in placement of the deposited material and provide in ease of disposition for many functional materials by using its' powder form. This proposed process shown in **Figure 15** is similar in working principle to laser welding and laser cladding in which using laser beam to form a melt pool and subsequently powder injected. The powder which is injected is deposited and fused onto the substrate because of the scanning from the laser. As the whole process is driven by the laser, so by controlling its beam and travel speed part can be manufacturing very accurately and it is become easy to mange other variables of the process and material waste as well as manufacturing time reduced [28].

4.6 Transformative manufacturing process

4.6.1 Sheet metal forming processes

Incremental forming process in one of the novel approaches of forming. Its basic working techniques shown in **Figure 16** is entirely different than traditional forming processes. A Single Point Incremental Forming (SPIF) uses a small tool which deformed the sheet metal and generate the shape of the final geometry. This process is different because in the process dies are not required to generate the shape. Below following set of relevant advantages is mentioned of SPIF [29]:

i. Process does not have any setup cost.

ii. A CNC is responsible for tool movement: a three-axis milling machine is sufficient for the requirement.

iii. Process offers very high flexibility: change in part program is very easy to do

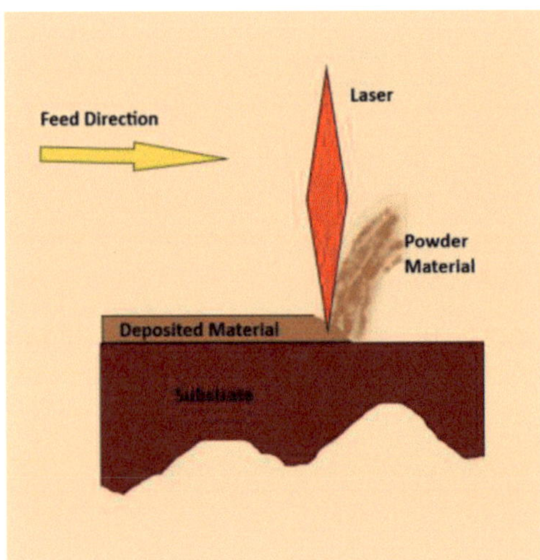

Figure 15.
Laser deposition system.

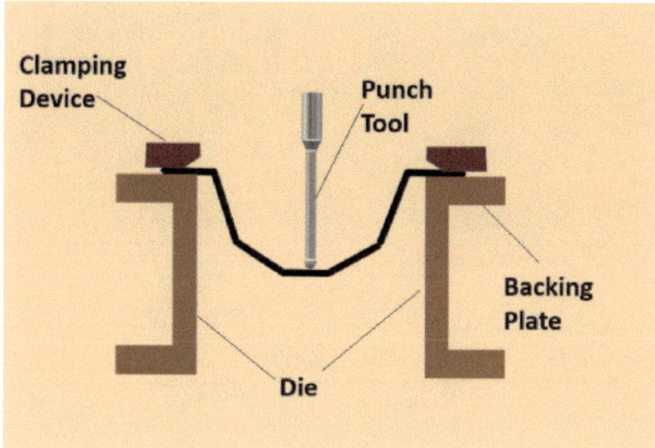

Figure 16.
Single point increment forming.

iv. The process is good alternative for rapid prototyping; the process is also very suitable to produce deform parts, especially old automobile body parts whose die is outdated and no longer available.

v. In comparison to conventional stamping processes, it provides a high formability. This characteristic of the process is due to the local deformation produced by the punch.

Some of the limitation of the product is listed below:

i. SPIF is comparatively slow process. The deformation is gradually done locally by the punch which travel on predefined ttrajectory to form complex shapes. Although advance machines operate at high tool feed, and it takes a time to form the sheet metal.

ii. The accuracy from the process is very limited. The effect of spring back can not be easily predicated, and operation need to be done on trial-and-error basis.

Incremented forming with die can be implied for specific requirement that is very hard to deformed by AISF. As one can see in the **Figure 17**, process has combination of two action AISF and SF. In result this process provides more uniform distribution of thickness. In both process stage AISF and SF induce same thickness across the sheet. After as per requirement less thinning can be generated by SF. If there is any pocket in the geometry it will help to form accordingly as material would be available to fill the pocket [30, 31].

4.6.2 Laser heat treatment and sheet metal forming

It is evident that the laser beam is provide a heat energy to change the microstructure and mechanical properties of the material, which result in ease of metal forming process. In the process shown in **Figure 18** a laser is utilized to heat sheet metal which eventually increasing the formability of the sheet and allows more effectiveness in single point incremental forming process SPIF. A laser beam is passed in front of the forming tool to heat the metal and to assist the forming process. A deep drawing process with laser assistance has been investigated. Prior to the deep drawing process energy of laser beam is used to heat the material locally, which allows the

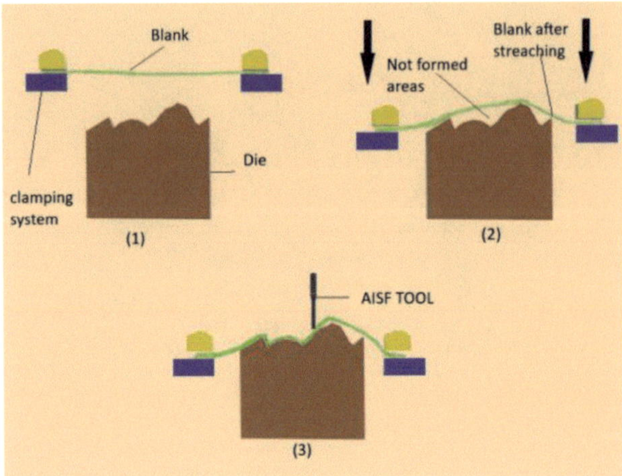

Figure 17.
Incremental forming with die.

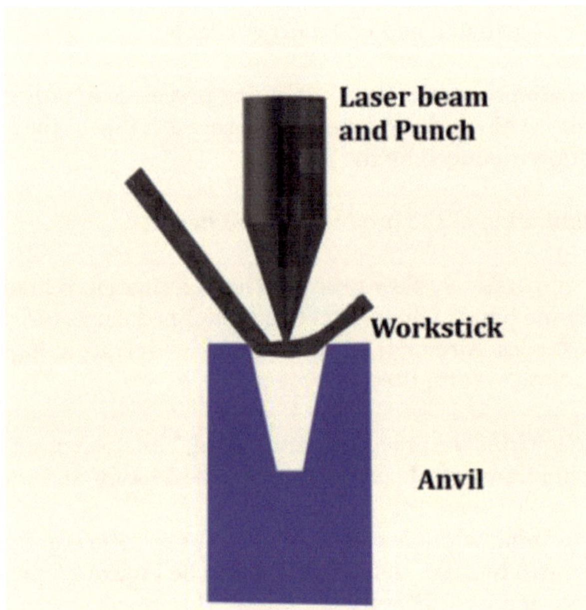

Figure 18.
Laser assisted V bending.

comparatively less drawing force, also it is responsible for lower number of forming steps and shows high productivity than the conventional deep drawing [32–35].

4.7 Subtractive manufacturing process

4.7.1 Mechanical machining and ECM

Figure 19 shows schematic view of the proposed hybrid process. The material removal process is driven by the combination of electrochemical machining and mechanical machining, a coated diamond abrasives spherical rod act as cathode and it rotates in pilot hole with certain speed in order removes material. The pilot hole is machined with a small diameter than the final hole diameter. In the process abrasives must be nonconductive while the tool core is electrically conductive. During

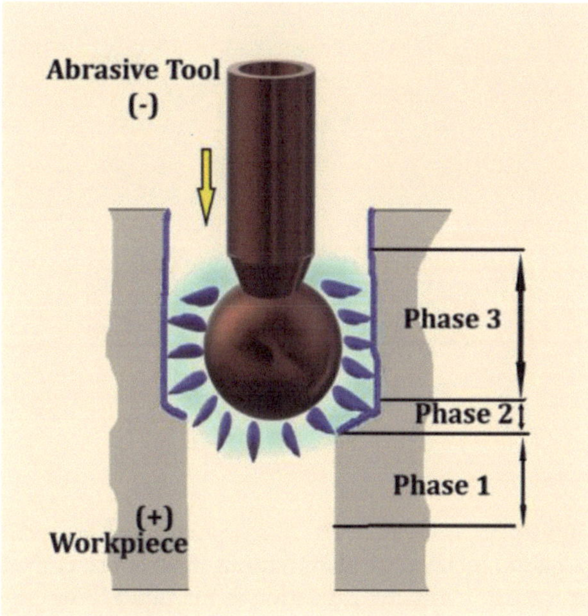

Figure 19.
ECM removing and grinding.

the operation, negative node is set on tool whereas positive end is set on workpiece. The diamond particles, which are abrasive particles of the tool are inserted before the nickel layer which is the conductive surface. This arrangement allows a gap between the conductive surface layer and wall inside the hole. In ECMG material removing action occurs in two stage, as it can be seen in **Figure 19**. Initially Electrolytic action start as soon as electrolyte filled the gap while during the process electrical charge supply to the tool. Step 1 of the process is entirely driven by electrochemical material removal action. Passive electrolyte NaNO3 is used in step 1 which allows passivation of the metal surface inside the hole. Step 2 of material removal process is a hybrid process which is combine action of electrochemical and mechanical material removal. As the tool advances inside the hole a gap between tool and material surface is decreases. The abrasive grains on the tool is responsible for removal of the soft and non-reactive passivation layer. In result of this step a fresh metal layer is become expose for another electrolytic reaction. During the process the electrolyte is stored between tools' diamond particles while the metal forms small cells of electrolytic, and that is how the dissolution of materials occurs. At the end of step 2 diameter of hole is enlarged and at its maximum limit. To manufacture highly accurate hole and sharp edges, insulated tool is used during the process. During step 3 only electrochemical dissolution occur which result in taper hole and no material removal is take place during the last stage [36].

4.7.2 Mechanical machining and EDM

In electrochemical grinding, demonstrate in **Figure 20**, allows material to remove by combination of abrasive action and electrochemical reaction; there are many variants of this process and abrasive wire ECM is one of them. Another variant of the process is allowing abrasive-laden air jet to directed towards the melt pool, and introduction laser milling/grooving processes has shown potential enhancement in the material removal rate (MRR) while almost eliminating the roughness and minimizing the heat affected zone of the generated surface. It is

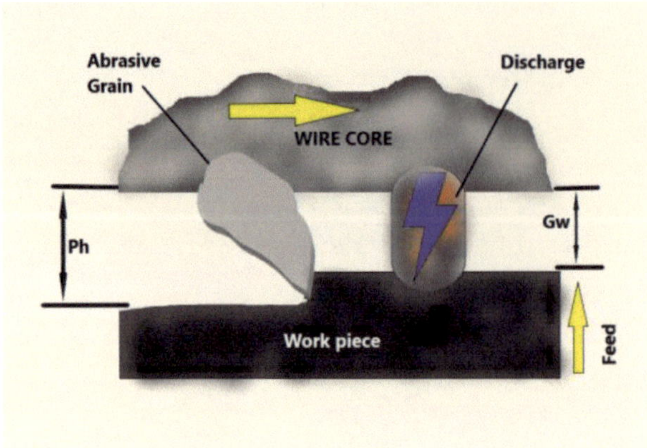

Figure 20.
EDM and mechanical machining.

evident that the hybridization in manufacturing processes is generally driven by the need that emerge due to technological limitations inherent to conventional manufacturing processes. Many hybridization in manufacturing processes applied abrasion removal machining process as one of the mechanisms by which material is removed. In electrochemical grinding process the combination of the abrasive action, which continually removes the surface material layers, and electrochemical material removal also contribute in removal of the material [37].

4.7.3 Laser jet and ECM

Electro chemical machining jet and laser drilling machining (JECM-LD) is not similar to laser drilling. It has combination of two different sources, energy of photons (laser drilling) and energy of ions (ECM), of energy simultaneously. The major purpose of combining a laser beam with a jet electrolyte is to achieve high quality machining by reducing the spatter and recast layer which is produced in simple laser drilling. As shown in **Figure 21** the focused laser beam is co-axially

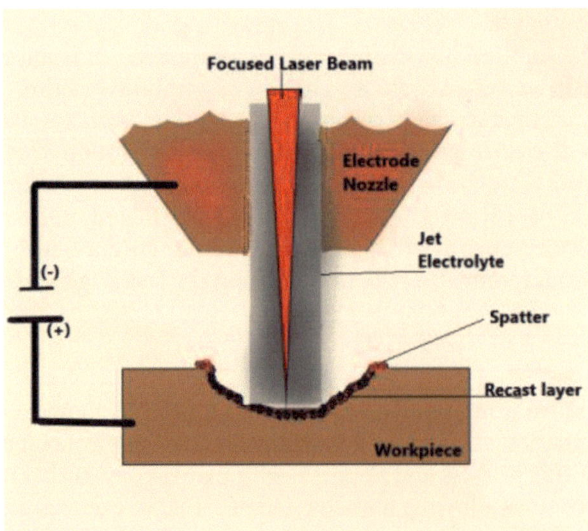

Figure 21.
Laser cutting and EDM.

aligned with a focused laser beam and tool this-electrode is not in contact with the material. The jet electrolyte and the laser beam are focused at a same time on the same location of workpiece. In JECM-LD manufacturing process, laser drilling is more responsible for material removing process and jet electrolyte effects is responsible to overcome the defects as it provides electrochemical reaction with materials, transporting debris and effective cooling to workpiece [38].

4.8 Ultrasonic assisted manufacturing process-(UAM)

Figure 22 illustrate the fundamental approach in ultrasonic assisted manufacturing processes and most potentially effective processes are mention below.

i. Ultrasonic assisted turning

ii. Ultrasonic assisted milling

iii. Ultrasonic assisted drilling

iv. Ultrasonic assisted grinding

v. Ultrasonic assisted EDM

vi. Ultrasonic assisted FSW

4.8.1 Ultrasonic assisted turning

Ultrasonic-assisted turning (UAT) is a novel approach of machining operation which generate a vibration by an ultrasonic system. **Figure 23** shows basic setup for UAT and tool behavior during cutting operation. The ultrasonic system generates high frequency and low-amplitude vibrations. Main purpose of this method is to avoid continuous contact with the workpiece. Most desirable benefits can be

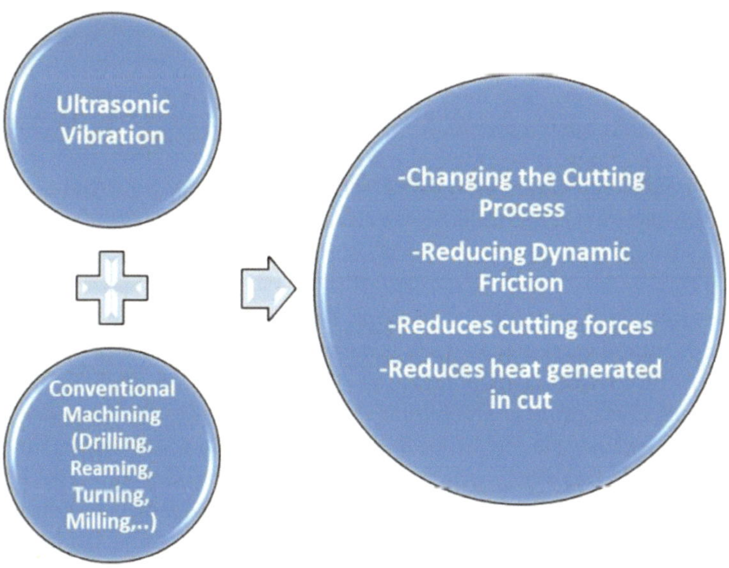

Figure 22.
Ultrasonic assisted process.

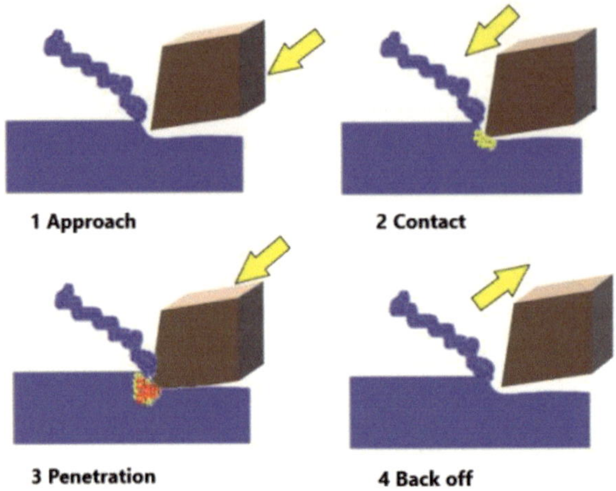

Figure 23.
Detailed view of ultra sonic turning process, 1–4 stages of the ultrasonic-assisted machining process.

achieved such as enhanced surface quality, reduced cutting forces, and lower residual stresses in a workpiece compare to conventional turning. Ultrasonic assisted is also advantageous for the machining equipment as it allows considerable life extension of the cutting tools because of lower requirements of cutting force. Ultrasonic Assisted Turning, however, is more efficient at lower cutting speeds compared to conventional turning. Ultrasonic-assisted machining effects decrease by increasing speed of cutting [39–43].

4.8.2 Ultrasonic assisted EDM

The combination of USM and EDM has the potential to reduce tool wear and electrode deflection in EDM of micro-holes and grooves. The mechanical signal was generated and transmitted to the tool-electrode, which was applied to remove material. The tungsten carbide workpiece was being vibrated while the EDM process was carried out. To remove the burrs formed on the exit region of a drilled hole ultrasonic vibration assisted dry electrical discharge machining process is used. The result of the study proves that in lower pulse durations the performance of novel ultrasonic vibration assisted dry EDM process is better compare to dry EDM process. The positive effects of applying ultrasonic vibrations to electrode at EDM process are because of both cavitation effect of the working fluid and also the vibrational action of the electrode itself [44–46].

Electrical discharge machining (EDM) is the process which is developed to remove conductive materials in the form of small craters. Such removal of materials are ranging from several to tens of microns. Process used electric sparks between a tool, electrode, and a workpiece which is submerged in a dielectric fluid. The sparks are strictly coordinated to control material removal rate. The micro-EDM process mechanism and EDM process mechanism is fundamentally same, the only notable differences are discharge energy supplied, tool dimensions and the resolution of the axis's movement. The gap between the tool and the electrode called spark gap, the series of sparks in a controlled spark gap is responsible for material removal, a small amount of material in the form of crater were removed from the workpiece as a spark strike the material. A pulse duration [μs], series of sparks within a certain time period, which is followed by a interval [μs], define as a pause for certain time duration in the sparking process. Each discharge cycle consists of pulse interval and

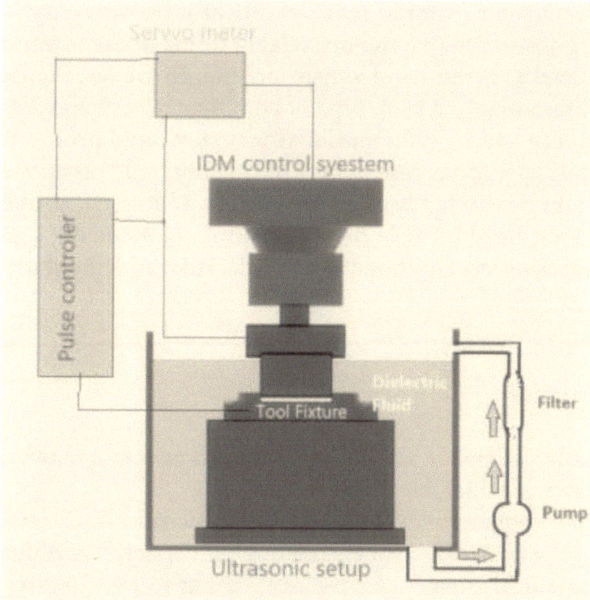

Figure 24.
Schematic diagram of ultrasonic assisted EDM.

pulse duration. The number of cycles which could run in a second control the discharge frequency. The pulse generator is used to control the discharge energy for a single pulse based on the requirement of machining conditions, such as finishing or semi-finishing, roughing. To maintain and to modify the spark gap and to control the movement of electrode respect to workpiece a servo control system is used. The EDM/micro-EDM system consist built in apparatus such as pump, filter, and dielectric reservoir to maintain the fresh dielectric flow in the spark gap which also flushing out the debris from the machined zone. **Figure 24** shows the basic sschematic diagram of ultrasonic assisted EDM/Micro-EDM system [45, 47–52].

5. Conclusions

It is observed tremendous amount of innovation and hybridization in advance manufacturing technologies. There are many researcher and universities are constantly working on innovative ideas and new technology. It can be observed that by hybridization of two manufacturing technology, more beneficial result can be achieved, and individual drawback of same process can be eliminated. Many of these innovations are potentially transformative, and not simply evolutionary. The subtractive processes and its combinations are mainly associated with the material, especially superalloys and ceramic, which are difficult to machine on the material removal processes such as milling, turning, drilling and grinding. The major contributors to material removal are EDM and other mechanical machining as such processes provided the high surface quality. Advance assisted processes such as ultrasonic vibration or laser cutting and its combination with conventional machining processes result in lower tool wear, higher surface integrity and shorter production times. Laser processing is still trending and attract many researchers to work on it in hybrid subtractive and transformative processes. It is important to be noted that the laser does not participate in actual materials removing process but introduction of it prior to the machining change the microstructures of the materials. It allows higher material removal rate as it become easy for conventional machining operations to remove the material in terms of the lower cutting forces which also beneficial as it results in longer tool life. However, flexibility of the processes is the limitation in a such type of combinations, therefore, to achieve freedom of flexibility and high dimensional accuracy rapid prototyping technology has been employed by various researchers to flexibly build components with arbitrary shapes. Future research advances, need to be addressed, namely, integration with other processes; need for new process-planning., modeling representations of hybrid process capabilities, additional standards, A.I. implementation (Machine learning) [44].

Acknowledgements

In the completion of this report, I have received encouragement and support from various quarters, which need special mention.

I wish to acknowledge my indebtedness to Dr. Parnika Shrivastava and Dr. Ramesh Guduru, Pandit Deendayal Petroleum University, Gandhinagar, who guided me throughout this work. And special thanks to Dr. Gülenay Kilic who support me at every step throughout my study, and for extending all possible cooperation.

I am also thankful to all those people of Pandit Deendayal Energy University, Gandhinagar, who helped me directly or indirectly during this period.

Mere acknowledgement may not redeem the debt i owe to my parents for their direct/indirect support during the entire course of this project.

Author details

Anand J. Patel[1] and Gülenay A. Kilic[2*]

1 Department of Mechanical Engineering, Pandit Deendayal Energy University, Gandhinagar, India

2 Department of Electric and Energy, University of Yalova, Turkiye

*Address all correspondence to: gulenay.kilic@yalova.edu.tr

References

[1] Z. Zhu, V. G. Dhokia, A. Nassehi, and S. T. Newman, "A review of hybrid manufacturing processes - State of the art and future perspectives," Int. J. Comput. Integr. Manuf., vol. 26, no. 7, pp. 596–615, 2013.

[2] K. P. Karunakaran, S. Suryakumar, V. Pushpa, and S. Akula, "Retrofitment of a CNC machine for hybrid layered manufacturing," Int. J. Adv. Manuf. Technol., vol. 45, no. 7–8, pp. 690–703, 2009.

[3] K. P. Karunakaran, S. Suryakumar, V. Pushpa, and S. Akula, "Low cost integration of additive and subtractive processes for hybrid layered manufacturing," Robot. Comput. Integr. Manuf., vol. 26, no. 5, pp. 490–499, 2010.

[4] K. P. Karunakaran, A. Sreenathbabu, and V. Pushpa, "Hybrid layered manufacturing: Direct rapid metal tool-making process," Proc. Inst. Mech. Eng. Part B J. Eng. Manuf., vol. 218, no. 12, pp. 1657–1665, 2004.

[5] X. Xinhong, Z. Haiou, W. Guilan, and W. Guoxian, "Hybrid plasma deposition and milling for an aeroengine double helix integral impeller made of superalloy," Robot. Comput. Integr. Manuf., vol. 26, no. 4, pp. 291–295, 2010.

[6] J. Y. Jeng and M. C. Lin, "Mold fabrication and modification using hybrid processes of selective laser cladding and milling," J. Mater. Process. Technol., vol. 110, no. 1, pp. 98–103, 2001.

[7] J. Zhang and F. Liou, "Adaptive slicing for a multi-axis laser aided manufacturing process," J. Mech. Des. Trans. ASME, vol. 126, no. 2, pp. 254–261, 2004.

[8] J. Ruan, K. Eiamsa-Ard, and F. W. Liou, "Automatic process planning and toolpath generation of a multiaxis hybrid manufacturing system," J. Manuf. Process., vol. 7, no. 1, pp. 57–68, 2005.

[9] F. Liou, K. Slattery, M. Kinsella, J. Newkirk, H. N. Chou, and R. Landers, "Applications of a hybrid manufacturing process for fabrication of metallic structures," Rapid Prototyp. J., vol. 13, no. 4, pp. 236–244, 2007.

[10] D. S. Choi *et al.*, "Development of a direct metal freeform fabrication technique using CO2 laser welding and milling technology," J. Mater. Process. Technol., vol. 113, no. 1–3, pp. 273–279, 2001.

[11] J. M. Pinilla and F. B. Prinz, "Lead-time reduction through flexible routing: Application to Shape Deposition Manufacturing," Int. J. Prod. Res., vol. 41, no. 13, pp. 2957–2973, 2003.

[12] A. M. Dollar, C. R. Wagner, and R. D. Howe, "Embedded sensors for biomimetic robotics via shape deposition manufacturing," Proc. First IEEE/RAS-EMBS Int. Conf. Biomed. Robot. Biomechatronics, 2006, BioRob 2006, vol. 2006, pp. 763–768, 2006.

[13] A. G. Cooper, S. Kang, J. W. Kietzman, F. B. Prinz, J. L. Lombardi, and L. E. Weiss, "Automated fabrication of complex molded parts using Mold Shape Deposition Manufacturing," Mater. Des., vol. 20, no. 2–3, pp. 83–89, 1999.

[14] M. Lanzetta and M. R. Cutkosky, "Shape deposition manufacturing of biologically inspired hierarchical microstructures," CIRP Ann. - Manuf. Technol., vol. 57, no. 1, pp. 231–234, 2008.

[15] A. Kelkar and B. Koc, "Geometric planning and analysis for hybrid re-configurable molding and machining process," Rapid Prototyp. J., vol. 14, no. 1, pp. 23–34, 2008.

[16] A. Kelkar, R. Nagi, and B. Koc, "Geometric algorithms for rapidly reconfigurable mold manufacturing of free-form objects," CAD Comput. Aided Des., vol. 37, no. 1, pp. 1–16, 2005.

[17] G. Lucchetta and R. Baesso, "Polymer Injection Forming (PIF) Of Thin-Walled Sheet Metal Parts — Preliminary Experimental Results, April 2006," , pp. 1046–1051, 2007.

[18] F. E. Pfefferkorn, Y. C. Shin, Y. Tian, and F. P. Incropera, "Laser-assisted machining of magnesia-partially-stabilized zirconia," J. Manuf. Sci. Eng. Trans. ASME, vol. 126, no. 1, pp. 42–51, 2004.

[19] H. Ding and Y. C. Shin, "Laser-assisted machining of hardened steel parts with surface integrity analysis," Int. J. Mach. Tools Manuf., vol. 50, no. 1, pp. 106–114, 2010.

[20] J. W. Novak, Y. C. Shin, and F. P. Incropera, "Assessment of Plasma Enhanced Machining for Improved Machinability of Inconel 718," J. Manuf. Sci. Eng. Trans. ASME, vol. 119, no. 1, pp. 125–129, 1997.

[21] M. C. Anderson and Y. C. Shin, "Laser-assisted machining of an austenitic stainless steel: P550," Proc. Inst. Mech. Eng. Part B J. Eng. Manuf., vol. 220, no. 12, pp. 2055–2067, 2006.

[22] E. Brinksmeier and T. Brockhoff, "Utilization of Grinding Heat as a New Heat Treatment Process," CIRP Ann. - Manuf. Technol., vol. 45, no. 1, pp. 283–286, 1996.

[23] K. Salonitis, T. Chondros, and G. Chryssolouris, "Grinding wheel effect in the grind-hardening process," Int. J. Adv. Manuf. Technol., vol. 38, no. 1–2, pp. 48–58, 2008.

[24] J. B. Taylor, D. R. Cormier, S. Joshi, and V. Venkataraman, "Contoured edge slice generation in rapid prototyping via 5-axis machining," Robot. Comput. Integr. Manuf., vol. 17, no. 1–2, pp. 13–18, 2001.

[25] H. Y. Liang, C. L. Kuo, and J. D. Huang, "Precise micro-assembly through an integration of micro-EDM and Nd-YAG," Int. J. Adv. Manuf. Technol., vol. 20, no. 6, pp. 454–458, 2002.

[26] H. Ou Zhang, Y. Ping Qian, and G. Lan Wang, "Study of rapid and direct thick coating deposition by hybrid plasma-laser manufacturing," Surf. Coatings Technol., vol. 201, no. 3–4, pp. 1739–1744, 2006.

[27] Y. Ping Qian, H. Ou Zhang, and G. Lan Wang, "Research of rapid and direct thick coatings deposition by hybrid plasma-laser," Appl. Surf. Sci., vol. 252, no. 18, pp. 6173–6178, 2006.

[28] J. Fessler, R. Merz, A. Nickel, F. Prinz, and L. Weiss, "Laser deposition of metals for shape deposition manufacturing," Solid Free. Fabr. Symp. Proceedings, Univ. Texas Austin, no. September, pp. 117–124, 1996.

[29] F. Micari, G. Ambrogio, and L. Filice, "Shape and dimensional accuracy in Single Point Incremental Forming: State of the art and future trends," J. Mater. Process. Technol., vol. 191, no. 1–3, pp. 390–395, 2007.

[30] L. Galdos *et al.*, "Enhancement of incremental sheet metal forming technology by means of stretch forming," AIP Conf. Proc., vol. 1315, pp. 601–606, 2010.

[31] B. T. Araghi, G. L. Manco, M. Bambach, and G. Hirt, "Investigation into a new hybrid forming process: Incremental sheet forming combined with stretch forming," CIRP Ann. - Manuf. Technol., vol. 58, no. 1, pp. 225–228, 2009.

[32] A. Kratky, "Laser assisted forming techniques," XVI Int. Symp. Gas Flow,

Chem. Lasers, High-Power Lasers, vol. 6346, p. 634615, 2006.

[33] R. Börner, S. Winkler, T. Junge, C. Titsch, A. Schubert, and W. G. Drossel, "Generation of functional surfaces by using a simulation tool for surface prediction and micro structuring of cold-working steel with ultrasonic vibration assisted face milling," J. Mater. Process. Technol., vol. 255, no. January, pp. 749–759, 2018.

[34] M. Geiger, M. Merklein, and M. Kerausch, "Finite element simulation of deep drawing of tailored heat treated blanks," CIRP Ann. - Manuf. Technol., vol. 53, no. 1, pp. 223–226, 2004.

[35] H. Shen, Y. Shi, Z. Yao, and J. Hu, "An analytical model for estimating deformation in laser forming," Comput. Mater. Sci., vol. 37, no. 4, pp. 593–598, 2006.

[36] D. Zhu, Y. B. Zeng, Z. Y. Xu, and X. Y. Zhang, "Precision machining of small holes by the hybrid process of electrochemical removal and grinding," CIRP Ann. - Manuf. Technol., vol. 60, no. 1, pp. 247–250, 2011.

[37] I. Menzies and P. Koshy, "Assessment of abrasion-assisted material removal in wire EDM," CIRP Ann. - Manuf. Technol., vol. 57, no. 1, pp. 195–198, 2008.

[38] H. Zhang, J. Xu, and J. Wang, "Investigation of a novel hybrid process of laser drilling assisted with jet electrochemical machining," Opt. Lasers Eng., vol. 47, no. 11, pp. 1242–1249, 2009.

[39] H. Puga, J. Grilo, F. J. Oliveira, R. F. Silva, and A. V. Girão, "Influence of external loading on the resonant frequency shift of ultrasonic assisted turning: numerical and experimental analysis," Int. J. Adv. Manuf. Technol., vol. 101, no. 9–12, pp. 2487–2496, 2019.

[40] Puga, Grilo, and Carneiro, "Ultrasonic Assisted Turning of Al alloys: Influence of Material Processing to Improve Surface Roughness," Surfaces, vol. 2, no. 2, pp. 326–335, 2019.

[41] P. Tandon and O. N. Sharma, "Experimental investigation into a new hybrid-forming process: Incremental stretch drawing," Proc. Inst. Mech. Eng. Part B J. Eng. Manuf., vol. 232, no. 3, pp. 475–486, 2018.

[42] M. A. Sofuoğlu, F. H. Çakır, S. Gürgen, S. Orak, and M. C. Kuşhan, "Experimental investigation of machining characteristics and chatter stability for Hastelloy-X with ultrasonic and hot turning," Int. J. Adv. Manuf. Technol., vol. 95, no. 1–4, pp. 83–97, 2018.

[43] M. A. Sofuoğlu, F. H. Çakır, S. Gürgen, S. Orak, and M. C. Kuşhan, "Numerical investigation of hot ultrasonic assisted turning of aviation alloys," J. Brazilian Soc. Mech. Sci. Eng., vol. 40, no. 3, pp. 1–12, 2018.

[44] S. Kumar, S. Grover, and R. S. Walia, "Analyzing and modeling the performance index of ultrasonic vibration assisted EDM using graph theory and matrix approach," Int. J. Interact. Des. Manuf., vol. 12, no. 1, pp. 225–242, 2018.

[45] N. Sabyrov, M. P. Jahan, A. Bilal, and A. Perveen, "Ultrasonic vibration assisted electro-discharge machining (EDM)-An overview," Materials (Basel)., vol. 12, no. 3, 2019.

[46] S. T. Kumaran, T. J. Ko, and R. Kurniawan, "Grey fuzzy optimization of ultrasonic-assisted EDM process parameters for deburring CFRP composites," Meas. J. Int. Meas. Confed., vol. 123, pp. 203–212, 2018.

[47] Y. Liu, H. Chang, W. Zhang, F. Ma, Z. Sha, and S. Zhang, "A simulation study of debris removal process in ultrasonic vibration assisted electrical discharge machining (EDM) of deep

holes," Micromachines, vol. 9, no. 8, 2018.

[48] R. Rajeswari and M. S. Shunmugam, "Comparison of Conventional, Powder Mixed, and Ultrasonic Assisted EDM by Phenomenological Reasoning," Int. J. Mater. Form. Mach. Process., vol. 5, no. 2, pp. 32–44, 2018.

[49] R. Rajeswari and M. S. Shunmugam, "Finishing performance of die-sinking EDM with ultrasonic vibration and powder addition through pulse train studies," Mach. Sci. Technol., vol. 0, no. 0, pp. 1–29, 2019.

[50] R. Rajeswari and M. S. Shunmugam, "Comparative evaluation of powder-mixed and ultrasonic-assisted rough die-sinking electrical discharge machining based on pulse characteristics," Proc. Inst. Mech. Eng. Part B J. Eng. Manuf., vol. 233, no. 14, pp. 2515–2530, 2019.

[51] M. R. Shabgard, A. Gholipoor, and M. Mohammadpourfard, "Numerical and experimental study of the effects of ultrasonic vibrations of tool on machining characteristics of EDM process," Int. J. Adv. Manuf. Technol., vol. 96, no. 5–8, pp. 2657–2669, 2018.

[52] P. Singh, V. Yadava, and A. Narayan, "Parametric study of ultrasonic-assisted hole sinking micro-EDM of titanium alloy," Int. J. Adv. Manuf. Technol., vol. 94, no. 5–8, pp. 2551–2562, 2018.

Chapter 7

Stress-Strain Relationship: Postulated Concept to Understand Genetic Mechanism Associated with a Seismic Event

Umesh Prasad Verma, Madhurendra Narain Sinha, Pushan Kumar Dutta and Subhra Mullick

Abstract

In this study, we propose the design methodology for monitoring the earthquake and for detecting and tracking micro-seismic changes in the earthquake prediction system. The alert device includes these sensors will be drastically different from current early warnings using the dozens of seismometers network across seismically active regions for measurement of small acceleration signals directly and, as the first, low-noise stage of the instruments measuring low-noise velocity signals. Strain develops over considerable time in the overlying stratum at right angle to the applied shearing (max) stress, obeying the internal friction of the stratum, available seismic energy and law of stress–strain relationship. Using estimated energy (seismic), stress accumulation, the addition or subtraction in the strain rate due to stress developed can be analyzed for a seismic event. This concept may lead to better understanding of stress generation; build up, transfer and final drop. Then we propose a methodology to identify type of data can be used for the spectral analysis in earthquake seismology and what type of instrument can be used for the spectral analysis in for data acquisition.

Keywords: Strain rate, Field velocity, Axis rotation, Stress-Strain relationship, Seismicity Stress generation

1. Introduction

Recent advances in seismic data acquisition techniques and equipment motivated the geo-scientific community in comprehending the complicated structure and dynamics of Earth, which has shown that plate tectonics can explain properly oceanic plate behavior having few large and rigid plates and but not fully well continental plates with models requiring quite a lot of small blocks due to in-situ material heterogeneity prevalent in deforming regions, such as rifts, spreading zones; collisional and subduction system including mountain belts. In this context, prediction of earthquake has become a very challenging task of scientists as it is not possible with the current state of knowledge of science because of involvement of complex physics of earthquake in its generating processes. It is the fact that we are trying to predict the unpredictable, which can be successfully achieved once the diagnostic parameters (precursors) responsible for genesis of the earthquake are known to us, which in turn suggests that the earthquake generating mechanisms are not completely known and well understood till today. In this proposed study, we review and revalidate the research aimed at unambiguously defined algorithms for seismogenesis that is central to the puzzle of geo-scientific community for mitigating the damage to floras and fauna for the benefit of entire mankind. The research conducted in this study deals with the integration of earthquake precursory analysis based on the integration of three methodologies, involving the phenomenological analysis of geo-scientific observations; universal models for the scientific validation of complex system design in statistical physics; and Earth-specific models of seismo-tectonic fault network observation. It is, therefore the study of earthquake, has become a subject of integrated multi-disciplinary geo-scientific research with aim of estimating a set of reliable diagnostic earthquake generating precursors so that prediction of the earthquake can be feasible with better degree of reliability.

This thesis is aimed at extending the scope of better understanding of earthquake generating mechanisms and its complexity in diverse geotectonic settings related to different seismogenic fault triggering systems (e.g., San Andreas Fault; East Japan fault; Himalayan Faults) for development of a holistic earthquake warning System for understanding earthquake generating processes in different tectonic environ, which can help us in adopting measures for mitigating earthquake hazards. It has been established that junctions of thrust faults and transverse lineaments near plate boundary are plausible vulnerable geotectonic settings for damaging earthquakes attributed to the rupture of the rock which occur in earthquake triggering zones following the accumulation of strain in the lithosphere. The complexity of the lithosphere shows that a multitude of mechanisms that affect the earthquake study are found to have certain space–time constraints. The motivation for defining a futuristic prediction scenario where rupture initiation and triggering models with various discontinuities is highly probable as a result of quasi-dynamic instability due to expansion of strain release areas and accumulation of strain in the fault patch has been critically examined in our study using nucleation models as a part of the design to define quasi-static nucleation models to explore the thrust fault locking zones in different earthquake prone regions. In order to enrich our understanding of the internal processes and deformation within the triggering zones that culminate into isolated fault segments with stress release and strain accumulation, we define certain interaction models which carry significant information about a short term precursory evaluation for the earthquake phenomena. The basic limitation of long term forecasting models and self organized criticality mechanism lies in its inability to define realistic geo-models able to explain the intricate forces involved in rupture based fault slippages on the nearby faults measured that may increase the overall stress on the linked fault in a short period of time. The earthquake genesis models can be defined mainly during short-term swarm events that cluster predominantly around the inside corner of ridge-transform intersections with events occurring on both the strike-slip and normal faults within the system. Precise earthquake locations along mid-ocean ridges, transform faults, and within hydrothermal systems can illuminate regions of active seismic deformation, and help us better understand the mechanics and kinematics of these plate boundaries. Earthquake prediction technique involves identifying the upper limit of strain for nucleation periods which is proportional to the magnitude of the mainshock. The nucleation patch is detectable everywhere locally in the earthquake causative fault and constrained the pre-rupture nucleation slip for destabilization of the stress-strength field due to realistic tectonic loading, spontaneous nucleation of the frictional instabilities and visco-elastic relaxation of the lithosphere. The modeled seismicity shows that nucleation mechanism leads to the non-reversible expansions of rocks and spreading of the ocean floor in the neighborhood of the rupture or along the weakest rocks during inter-plate earthquake cycle. Our study starts from defining the problem of earthquake forecasting in the context of quasi-static dynamical models for earthquake nucleation initiation and subsequent arrest in the fault patch that is incorporated based on the time- and stress-dependencies of the nucleation process namely pre-slip models, colliding clusters or coalescence models, dilatancy models, characteristic slip or fixed time recurrence models and spatio-temporal clustering models where recurrent slip occurs in time with aftershocks, foreshocks, and pairing of main shocks. The present study proves that only way we can define earthquake forecast is through analysis of the source of earthquakes through interdisciplinary approaches of seismo-tectonic studies by observing strain accumulation and investigation of the physical change and temporal change of the rock properties in the earthquake source region. The variations of the stress field precede the main shock by days up to months through a sequence of simple accelerating nucleation process, and clearly inter-event triggering and a pre-existing near-critical stress field, operating over much longer correlation length scales in the incipient region called triggering basin which has been identified using spatial ant colony optimization models. Evolutionary robust computational algorithms are found to help in the recognition and understanding of the triggering mechanisms for localization of the focal area in the foreland basin in a near-critical state condition.

In the scheme for data analysis and earthquake genesis model section, we have developed and tested methods analyzing low signal to noise non-stationary data with quasi-periodic components characteristic of earthquakes. The dependence of seismicity on the general properties of the fault network can help in the analysis of local fracture stability by analyzing common characteristics of the seismicity behavior, identification of the missing attribute for seismicity and through identification of the unobserved models of seismicity which are characterized in the time series data. We focused on different data based techniques involving continuous time signals for earthquake analysis such as catalogs, acceleration-time series data, radon count analysis, image-processing techniques that can be used in validating and testing earthquake models. In this study, several case studies derived from catalogs for earthquakes occurring in the Himalayan Basin from about six decades of different researchers are taken up for analyzing the seismic trend using least square regression and weighted least-square regression. Local search models can be applied for analyzing rupturing and interaction for earthquake nucleation and triggering mechanisms can be quantified and their correlation can be studied with feed-forward neural network with backward projection for studying spatio-temporal seismicity clusters in the Himalayan Basin based on the physical processes governing dynamic rupture nucleation, growth, and arrest. The heterogeneity of the strength distribution for a recurrent time –asperity model has been found to depend on the critical strain as crack heads nucleate in meta stable condition and numbers increases rapidly. In this regime dominated by nucleation and growth of individual fractures, the population law exhibits power length distributions as the nucleation rate decreases as more regions become part of the stress affected regions resulting in the decrease in the number of fractures. This has been proved in the Himalayas subduction region could be explained by a coupling process between the observed physical parameters and the earthquake preparation processes. We have also used a Kalman particle filter and an improved Periodogram analysis in the design of a novel algorithm to find the statistical estimators for extracting the cumulative slip from the earthquake acceleration data for the 2011 Sikkim Earthquake Data. In this approach, an attempt is made to apply different algorithms to undertake detailed study on thrust-fault behaviour for better understanding of earthquake generating processes. A new algorithm is proposed to estimate rupture directivity using peak ground acceleration data instead of strong motion data, and we found that peak ground acceleration parameter with rupture directivity is an important input with immense potential for earthquake forecast and warning system.

In this study various conceptual models are analyzed among which a quasi-static nucleation model is used for validating the seismogenesis of earthquakes occurred in different tectonic settings. The fragmentation of structures and the rotation of the blocks is found to affect the driving force analyzed through non-linear correlation models which impart information based on the temporalspatial evolution processes of fault stress in the stages of stress deviating from linearity and meta instability in two or three dimensions as colliding cascade, pre-slip, characteristic earthquake models, dilatancy model and spatio-temporal clustering of earthquake models. The dilatants model has shown that fault edges can stay locked over an infinite depth over the entire seismic cycle, possibly by volumetric deformation in the seismogenic zone at different stages of the seismic cycle. The connection between seismicity and geodynamic models has been found rejuvenated using Radon emission in the San Andreas Fault from 1978 to 1981 and the Bayesian likelihood for a drastic change in the strength of an array and reset the elastic stresses by external forces (earthquake) occurs when certain changes in the structure of the array as the compressive and tensile strength of the rock array. Based on subjective analysis we analyze how this model enable or inhibit rupture to overcome regions of faults unfavorable to rupture.

In this piece of research, it is found that there is a need to reliably describe the complexity of the source of a seismic event by evaluating parameters pertaining to changes in stress, strain and rheology of seismic deformation processes apply discrete-time Hidden Markov Model in order to reveal the stress field underlying the earthquake generation. For the Tohoku earthquake, we have studied fault segmentation through studying fault inter-segment areas usually associated with local slip and fractal clustering through Graph cut based image transformation carried out for pre and post earthquake image analysis applied for the 2011 Tohoku earthquake for satellite image analysis. In the later meta-instability stages, it has been found that strain release areas expand by linking with each other indicating the instability leading to an impending earthquake event based on the dynamic variation associated with meta-instability stage. The increase of such areas is a practical indicator of the expansion, accelerated expansion and linkage of nucleation models that give us a tested approach for earthquake forecasting called System of geodynamic monitoring in localness, radon behavior models and rotation of the earth to observe the difference in the rate of acceleration or de-acceleration of the tectonic plate prior to a mega-thrust earthquake using earth rotation data for the 2011 Tohoku earthquake for different time events to relate seismicity with the Earth spin changes. The variations of the stress field precede the main shock by days up to months trigger the role of faults in regional deformation in the time of around 150 days. The recognition and understanding of the triggering mechanisms can help to the localization of the focal area in the foreland basin. Triggering basins serve as harbingers of large earthquake where stress–strain interactions have been analyzed by the quasi-static mechanics of seismic precursory stress–strain propagation in the crustal lithosphere that can make time-dependant rupture analysis and define the likelihood of the occurrence of the earthquake in the future [1]. The first stage of the rock fracture begins when the stress curve deviates from linearity as differential strain variations occur at every portions of the fault due to change of rock stiffness equivalent to the tensibility co-efficient resulting in isolated areas of stress release and strain accumulation. In the second stage, strain release areas associated with quasi-static instability undergo a state of locking that initiates the early meta-instability for non causal behavior of stress variability that initiates instable slip of fault as an independent activity for dissipation in nucleation of each fault part into a steady state evolution for stress transfer [2]. When the unstable slip function associates itself to the attractor states for the optimal state sequences, the fault enters the meta instability stage of the rock matrix when isolated strain release areas of the fault increase and stable expansion proceeds as a distribution of the residual strain and is augmented by the sub-surface heterogeneous environment. All these observations suggest that understanding of the earthquake generating processes requires multi-disciplinary approach which can be analyzed based on the rock properties using a reduced roughest approach for rock characterization. When slips occur between the crust and the tectonic plate, the stored elastic energies are released in "bursts" which can be detected as earthquakes. As the stresses become more intense, the elastic strain energy stored in a portion of the crust (block), moving with the plate relative to a "stationary" [3] neighboring portion of the crust, can vary only due to the random strength of the solid–solid friction between the crust and the plate. This may suggest clustering and specifically localization around planes, migration, spatio-temporal gradients of seismic parameters within a limited range. As the steady expansion is associated with quasi-dynamic instability, interaction between the areas expands with fault linking as earthquakes are generated. The process of meta-stability shows that weak and strong segments associated with fault strain release and strain accumulation is found to occur successively as relatively independent and quasi-static triggering of the rock. The location and timing can be identified for linking the segment and conducting a location error based analysis for optimal allocation of wireless sensor networks nodes which measure the received signal strength for a certain parameter that is highly useful in designing for the earthquake early warning system (EEWS) in future. Our comprehensive study made us understand that the reliability of the EEWS depends up on the degree of integration among various methodologies and models, which have strong bearing on determination of earthquake location, depth, size and the nature and extent of fault rupturing.

Stress generation effect and its mechanism causing seismic forces[1] [4, 5] related with endogenous and exogenesis sources both. Correlation of data on strain rate for past events of Tibet, Anatolia, 1994, Sikkim, 2011 and Turkey, 2011 supports the concept. Astrophysical like (celestial objects viz.; Sun, Moon, mars. and Jupiter) influence on gravitational pull on the Earth which ultimately is responsible for the stress building forces at the interface of Lr-up mantle[2] [5, 6] affecting seismicity at an area. The nucleation patch is detectable everywhere locally in the earthquake causative fault and constrained the pre-rupture nucleation slip for destabilization of the stress-strength field due to realistic tectonic loading spontaneous nucleation of the frictional instabilities and visco-elastic relaxation of the lithosphere. It is necessary to identify the duration for earthquake rupture initiation for the dynamic stress drop and episodic changes involved for the path of stress relaxation to the most stable condition. The strain rate measurement by GSRM[3] [5, 9], and GPS data

[1] Stress generating forces are consequences of either exogenetic or Endogenetic effect as astrocelestial neration is initiaobjects' dynamics and status of grvatatonal forces.

[2] At the interface of Lr Up mantele interface stress generation is initiated due to reactive force of gravtational pull interacting in between sun –moon and earth system.

[3] GSRM stands for the Global strain rate measurement data which in elaborate manner has been produced by zhao jeng and E Hot of Stony Brook universitymore in formationin reference ctation no [5, 7, 8] can be available.

system[3] [1, 2, 7, 10–12] have cited theirs work on the strain rate measurement on the Asian region of Tibet and Anatolian plateau. Pattern of strain rate expansion and contraction and calculation of axis rotation and field velocity put interesting observations to be investigated in their works. Contraction alludes for overlooking the impending seismicity and expansion for positive occurrence. Through the equation below

$$\varepsilon_{xyz} = \frac{1}{\varepsilon(\sigma_3 - \sigma_3)} \propto C\Delta T \qquad (1)$$

ε_{xyz} is utilized for measuring the 3D strain development due to applied stress on the reservoir rock underlying the stratum. On real time analysis with the available data on strain rate and field velocity measurement: Kreemer et al. [13] and Marrett and Allmendinger [14] could not establish the concrete relationship between stress and strain behavior to follow up for an impending event. In the present paper mission is to encounter the shortcomings. The latest model version of May 2004 (i.e., GSRM version 1.2) includes 5170 velocities for 4214 sites worldwide [11]. The model consists of 25 rigid spherical plates and ~25,000 0.6° by 0.5° deformable grid areas within the diffuse plate boundary zones (e.g., western North America, central Asia, Observations on the available data set for tensor moment of earthquake events since 1976 to recent period on Asian region Viz Anatolia, and Tibetan plateau[4] [8, 16] shows haplessness to foretell the forthcoming events arrival due Strain rate measurements, over an area by GSRM[4] [8] and GPS data sytems[4] [8, 15] help in establishing the nature and pattern of the medium and strain velocity. Expansion and contraction rate details, at times, impede detection of exact magnitude, direction of vectors and axis rotation. Minute measurement of direction and magnitude of strain rate, using the proposed concept, helps in better understanding of seismicity.

The proposed work is used to highlight the novel concepts developed through investigation to explain the appropriate mechanism and cause of seismicity at an area on the basis of strain velocity vectors (in expansion and contraction term) The findings of facts are in support by the ISAR Infermetry; Price and Sandwell [17] for

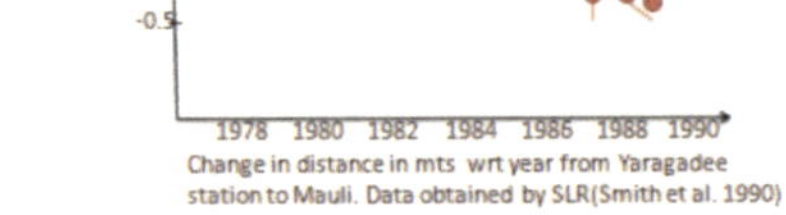

Figure 1.
Principle and mechanism: Principal stress on the rock volume considered cause the strain following law and rules of rock mechanics[6] [7, 9] first, the Collaboratory study of earthquake predictability (CSEP; Jordan et al. *[21]) is accepting global agreement in the seismology.*

[4] Global positioning data reveals besides the GSRM data for strain rate vectors size and direction for Tibbetan and anatolia plateau which are insufficient to caculate itself stress generation on the site. In Ref. [8, 15] comprehensive information are laid for pattern and medium of strain vectors progress.

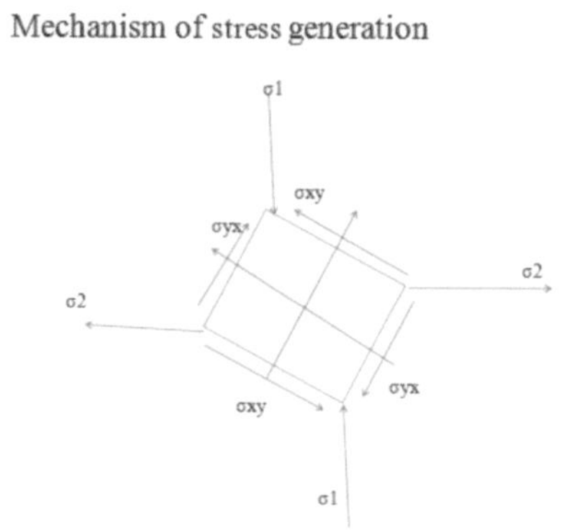

Figure 2.
An attempt to explain the nature of strain produced to impending principal (devitoric) stresses in the horizontal plane over the stratum rock unit: Courtesy: Fundamentals of [19].

event of Landers, California Earthquake (7.3Mw) on June 28, 1992.[5] [18–20] (**Figures 1** and **2**).

Expression from [19][6] [18], we have equation below.

$$\varepsilon_1 = \frac{1+v}{\varepsilon}\{\sigma_1(1-\gamma)v\sigma_2\} \tag{2}$$

Which was put in by Smith et al. [22][6] to be cited from Refs. [9, 23].

Symbols has meaning like: ϵ = strain; ν = possion ratio; o1, σ2 are (deviatoric) principal (deviatoric) stresses; ε = coefficient of elasticity of the materials interacted by stress.

Again, strain the y axis direction[7] [4, 7, 10].

$$\varepsilon_2 = -\frac{1+v}{\varepsilon}\{\sigma_2(1-\gamma)v\sigma_1\} \tag{3}$$

With the thermal stress effect on the stratum rock thermally unconstrained we get change in temp. That can lead to large change in pressure or stress as in the equation[8] [4, 7, 10].

---

[5] ISAR Infermetry; Price and Sandwell [17] for event of Landers, California Earthquake (7.3 Mw) on June 28, 1992)[9] [18, 19]. Inverse Synthetic aperturer Radar system on real time sampling is available by Phillipe Laccomme, Eric Norment in Air and spaceborn Radar system,2001.:Sc Direct topic, http// Science Direct .com landrs california earthquake (7.3.Mw) June, 28 in1992 was worked out with the ISAR infermetory dta analysis by Price and Sandwll [17]. It includes principles of Rock Mechanics and in seismology. These rules use mathematical modeling and equations of stress strain relationship in context with viscosity and linear expansion in the rock body due to thermal stress and produced strain symptomised by strain field vectors. Discussed in the reference [7, 9] as predictibility solution for any seismic event by Jordan et al. (2007)

[6] Expression from the rock mechanics [18, 19] we have, in the Eq. (2) of the text stated. Mathematics and relationship in between strain rate Є1 elasticity ε of rock interacting stress σ1and viscosity ν with thermal strain Ý are revealing direction and amount of strain prouced over the rock surface. These are viewable in form of vectors progression over the GSRM map by E Holt.

[7] Smith et al. [22] have implied the use of Eq. (2) in his literature Refs. [9, 23].

[8] Similarly as in Eq. (2) stress σ2 interacts as intermediate principal stress along horizontal direction yielding strain at right angle to the imposed direction as Є2 in Eq. (3), whose better explanation is available in texts of Refs. [4, 7, 10].

$$\varepsilon_1 = \frac{1}{\varepsilon}\{\sigma_1 - v\sigma_2\}\alpha xT \quad (4)$$

From the theory of linear elasticity, Muskhelishvili [3].[9] Further linear elasticity is modified as[10] [9].

$$\varepsilon_1 = \varepsilon_2 = \varepsilon_3 = -\frac{1}{\varepsilon}\alpha v \Delta T \quad (5)$$

And[10] [9],

$$\varepsilon_1 = \frac{1}{\varepsilon_1}(\sigma_1 - v\sigma_2 - v\sigma_3) - \alpha 1 \Delta T \quad (6)$$

$$\varepsilon_2 = \frac{1}{\varepsilon_2}(\sigma_2 - v\sigma_3 - v\sigma_1) - \alpha l \Delta T \quad (7)$$

$$\varepsilon_3 = \frac{1}{\varepsilon_3}(\sigma_3 - v\sigma_2 - v\sigma_1) - \alpha l \Delta T \quad (8)$$

These Eqs. (7)[11] [9] and (8)[12] [9] will give associated effect of temperature and thermal expansion over the Rock Staratum.

In vertical direction for half space, σ_3 is 0 and[12] [9],

$$\sigma_1 = \frac{\varepsilon \alpha l \Delta T}{1 - v}\sigma_2 \quad (9)$$

From the Text of Rock Mechanics substituting these values into (4) we obtain

$$T = \Delta T = \exp\frac{y\sqrt{\omega}}{2k} + \cos\left(\omega t - y\frac{\sqrt{\omega}}{2k}\right) \quad (10)$$

Where y = surface length, k = thermal diffusibility in m2/s K, c = specific heat of the rock of stratum, α = linear coefficient of expansion. $\sigma_{\max} = \frac{\varepsilon \alpha l \Delta T}{1-v}$ With thermal and elastic stress effect on rock stratum[13] [9, 17].

---

9 Eq. (3) states the Absolute value of strain amount as vector size on map in X direction as consequent interaction of Principal stress in x direction minus intermediate stress $\sigma 2$ in y axis multiplied by possion ratio and mnus joint effect of temperature, thermal expansion and poission ratio of the rock under stress.

10 From the theory of linear elasticity, by Muskhelishvili (1963) which is to be cited from Ref. [9] It. He is briefed as in the referenced author strain produced over a rock in all the three x,y and z direction are equal to one third of linear expansion × possion ratio × tthermalgradient as in following Eq. (5) Further linear elasticity is modified in Ref. [9].

11 In the Eq. (6) of manuscript value of strain depends on the principal grwates stress subtracted from joint effect of intermediate and least stress in the space due to convestive thermal activity which has been dealt in Ref. [9].

12 Eqs. (7) and (8) of text is incorporating the strain value in y and z direction respectively as per interacting stress secondary and r (intermediate 0 and least acting at perpendicular to the strain production direction and is cited from the work from Ref. [9].

13 Deal with the stress calculation steps in the Eqs. (10) and (11) involving frequency and phase of wave (of stress) due to thermal difusibility and thermal expansion of rock, its elasticity and possion ratio whose relationship is output from Rock Mechanics cited in Refs. [9, 17].

$$\sigma_1 = \sigma_2 - \frac{h}{1-v}(\rho g v + \varepsilon \alpha l \beta) \tag{11}$$

Here β is thermal gradient; With thermal and elastic stress effect on rock stratum: Pressure at depth h is[13] [4, 7–10, 16–18, 23].

$$\mathrm{P} = \mathrm{P} = \mathrm{P} = \frac{1}{3}(\sigma_1 + \sigma_2 + \sigma_3) = \frac{1}{3}\frac{1+v}{1-v}\left(hpg + \frac{2}{3(1-v)}\varepsilon \alpha l \beta\right) \tag{12}$$

Thus from the mathematical equations derived from the rock mechanics text and elasticity theory strain behavior can be understood. It is depending on the thermal stress condition and related parametric sources to be developed in possible direction.

Citing Sources: GSRM data from EHolt and A (Zhao (1997–2002)[14] [7, 9, 17, 23].

Figure 8 Above states the nature of strain against impending stress on the quartzite rocks observed by Bienweiski [24] we infer from the stress: Strain relationship whatever the type of stress Litho static-or devitoric shearing (axial) or normal the effect on the rock surface of the stratum[15,16] [8, 9, 16, 17].

2. Data acquisition

Figure 3 is the strain rate measurement involving GSRM map from E Holt and Zhao Zheng.

Figure 4 displays the 25000 grids distribution on the global map. Entire network of strain rate or velocity vectors shows expansion and contraction due to upwelling stress from the (mantle-up-lr) interface. Interesting point is that most of velocity

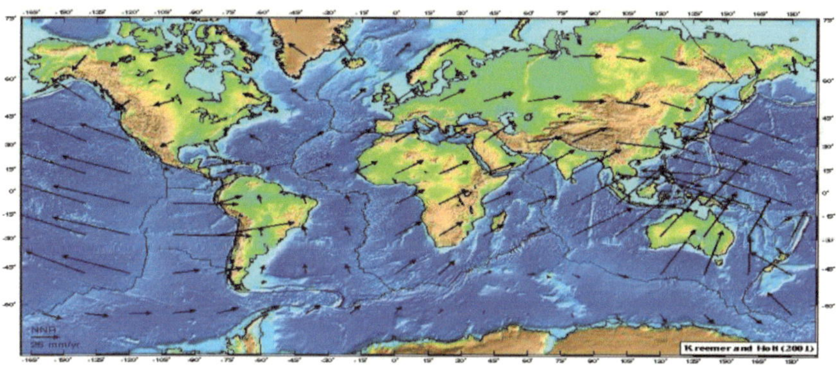

Figure 3.
Courtesy; E.Holt (2004) for strain rate measurement on GSRM.

[14] In the Eq. (12) pressure over rock body is calculated as the sum of stress due to thermal convestive current in all three x,yaxxes direction and with joint effect of possion ratio and thermal gradient and expansion coefficient is attributed.. as per reference dalt in [4, 7–10, 16–18, 23].

[15] E Holt and A Zhao (1997–2002) have worked over assembling global strain Rate measurement data in form of vectorial notations in size and direction wise as in the **Figures 2-8** which have been discussed in the Refs. [7, 9, 17, 23] of bibliography.

[16] Rock stsrata experience stress interaction in upward progression of deviotoric (shearing) or confining (Lithostatic stress).it has been detailed by Bienweiski, [24] in his literature for stress behaviour over Quartzite formation. Can be referred in [8]. Elastic anisotropy of regularly jointed media [18].

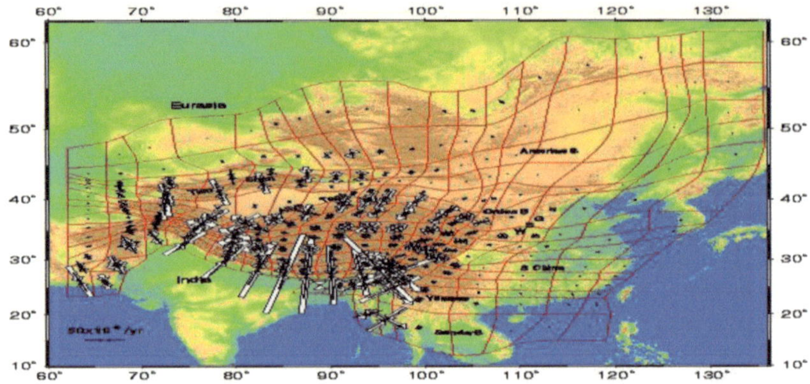

Figure 4.
Global Strain Rate Measurement courtsey Zhao.

vectors on the Asian continental plates are in contracting mode. Vectors at the oceanic plates are longer and in expanding mode. It is obscure about the stress stored below the crustal reservoir having contracting mode and stress energy already been consumed at the expanding region. It has been significantly observed during the 18th September, 2011 outbreak at Sikkim (6.8 Mw) and Turkey 7.2 Mw on October 24th 2011.

In the **Figure 5** velocity vectors are due north and NE on Indian subcontinent and Tibetan plateau region, respectively. Vectors show the Indian plate moving towards the Tibetan plate. Eurasian plate remains passive w.r.t. Pacific and American Plates. It has been evident even by the Harvard CMT catalog data set. From Jan 1977 to November 2002.

Figure 6 shows the pattern of strain rate or vectors with greater in amount than the vectors at continental plates of America with respect to Pacific. Eventually has been observed y the events of Chile 7.3 Mw on 2nd Jan 2011 and Mexico 6.5 Mw on October 2011. Vectors are shorter at the continental region than that of Oceanic region of pacific. This has been evident even by the Harvard CMT data catalog since Jan 1977 to Nov 2002 (**Figure** 7).

Figure 8 above states the nature of strain against impending stress on the quartzite rocks observed by Bienweiski [24] we infer from the stress–strain relationship whatever the type of stress Litho static-or devitoric shearing (axial) or normal the effect on the rock surface of the stratum depends on the parameters.

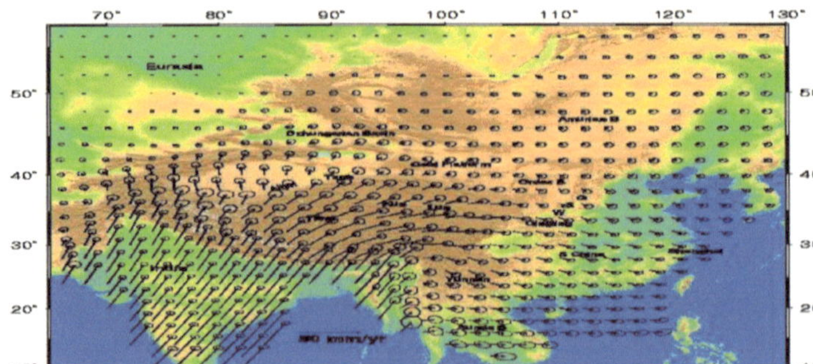

Figure 5.
The supportive evidence of concept laid in the abstract.: Courtesy: Zhao et al. Kreemer et al. [25] strain rate measurement for the Tibetan and Asian region prior to Sikkim (2004) 5.7Mw event.

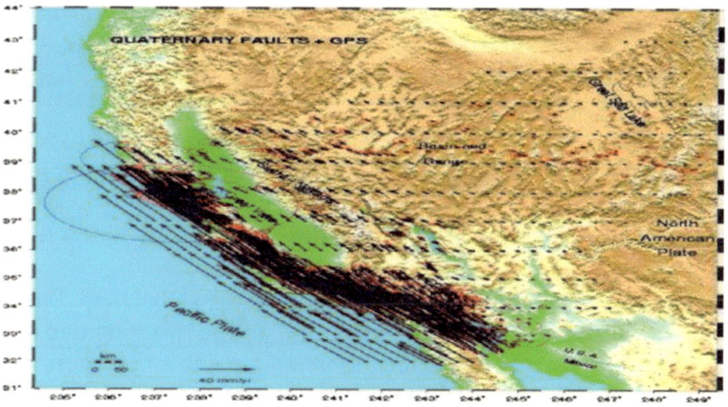

Figure 6.
Velocity vectors of Pacific plate against American fixed plate(relatively) courtesy: Almindger Zhao et al. (1998, 2004).

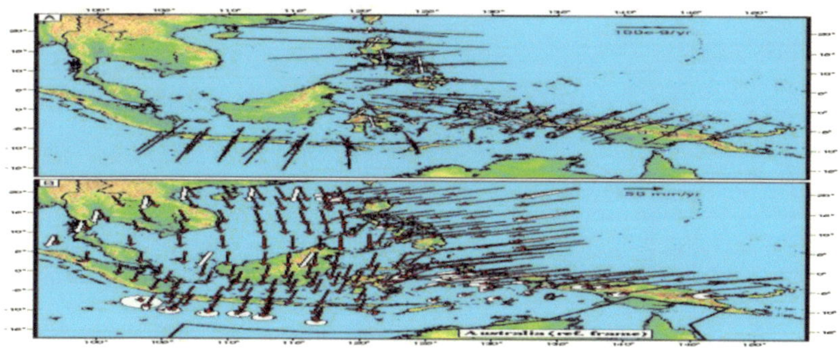

Figure 7.
The direction and amount of vectors in the oceanic region of Indian and pacific plates with relative passiveness of Asian plate.

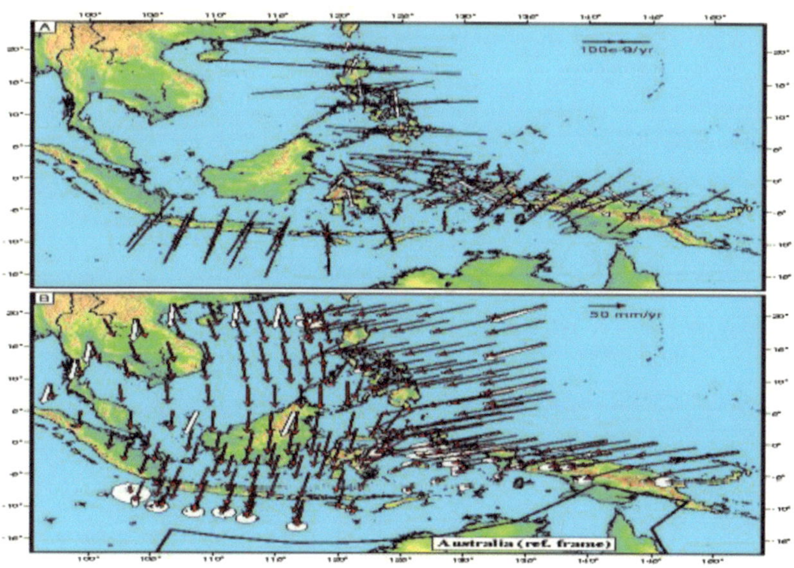

Figure 8.
Courtesy; by E. Holt (2004) for strain rate measurement on GSRM.

Characters like rigidity G, elasticity ε, density ρ, Poisson ratio υ coefficient of thermal expansion α, thermal diffusibility β.

Figure 7; ISAR Infermetry; Price and Sandwell [17] for event of Landers, California Earthquake (7.3 Mw) on June 28, 1992 [18, 19].

Figure 2 is an attempt to explain the nature of strain produced to impending principal (devitoric) stresses in the horizontal plane over the stratum rock unit (**Figures 8** and **9**).

Figure 3 is the data acquired from the GSRM map worked out by E Holt and zhaon Zeng 1979 et al.

Figure 4 displays the 25000 grids distribution on the global map. Entire network of strain rate or velocity vectors shows expansion and contraction due to upwelling stress from the (mantle-up-lr) interface. Interesting point is that most of velocity vectors on the Asian continental plates are in contracting mode. Vectors at the (**Figures 10** and **11**) shows the application frequency range of the circuit to replace the inductance.

Figure 12: velocity vectors of Pacific plate against American fixed plate (relatively) courtesy: Almindger Zhao et al. (1998, 2004) **Figure 6** shows the pattern of strain rate or vectors with greater in amount than the vectors at continental plates of

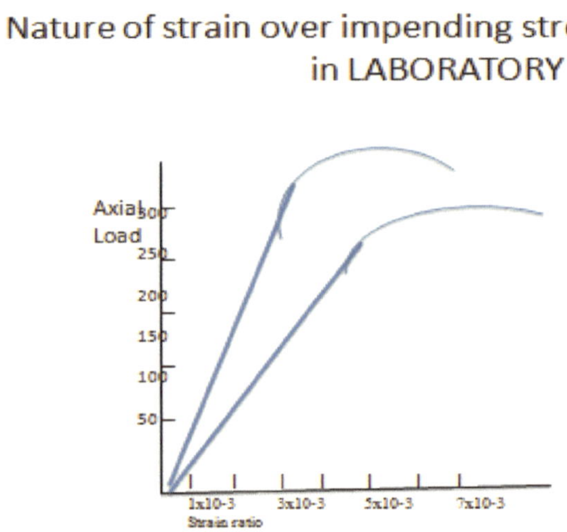

Figure 9.
Courtesy: Fundamentals of [19].

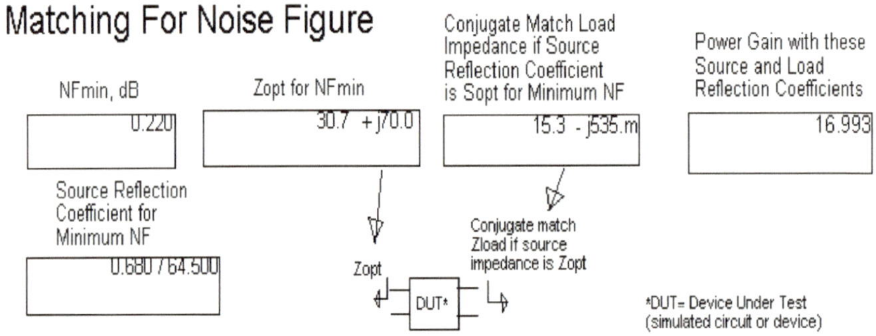

Figure 10.
ATF34143 best noise matching conditions.

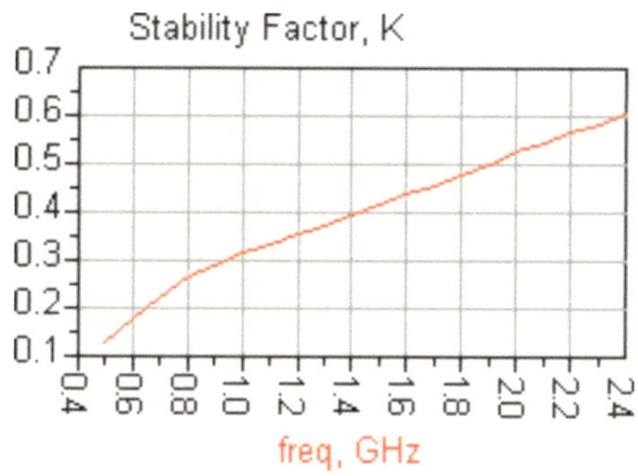

Figure 11.
ATF34143 stability analysis results. (a) Increase the inductance at the source to make the circuit stability greater than 1 in the application frequency range, and then replace the inductance with a microstrip line.

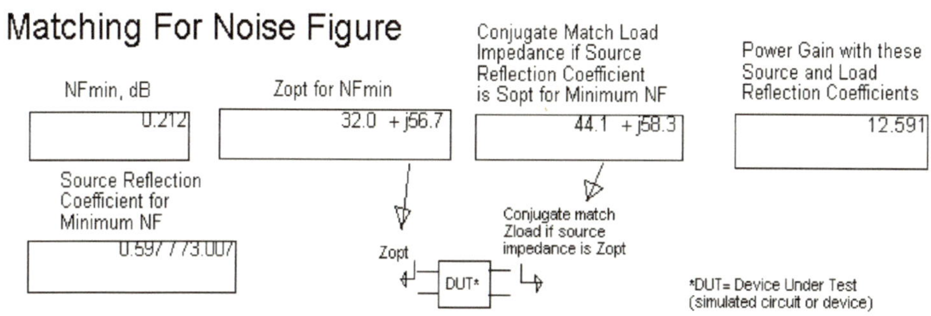

Figure 12.
ATF34143 the best noise matching under the condition of increasing source inductance.

America with respect to Pacific. Eventually has been observed y the events of Chile 7.3 Mw on 2nd Jan 2011 and Mexico 6.5 Mw on October 2011.vectors are shorter at the continental region than that of Oceanic region of pacific. This has been evident even by the continental plates of America with respect to Pacific. Eventually has been observed y the events of Chile 7.3 Mw on 2nd Jan 2011 and Mexico 6.5 Mw on October 2011.vectors are shorter at the continental region than that of Oceanic region of pacific. This has been evident even by the Harvard CMT data catalog since Jan 1977 to Nov 2002 (**Figures 13–16**).

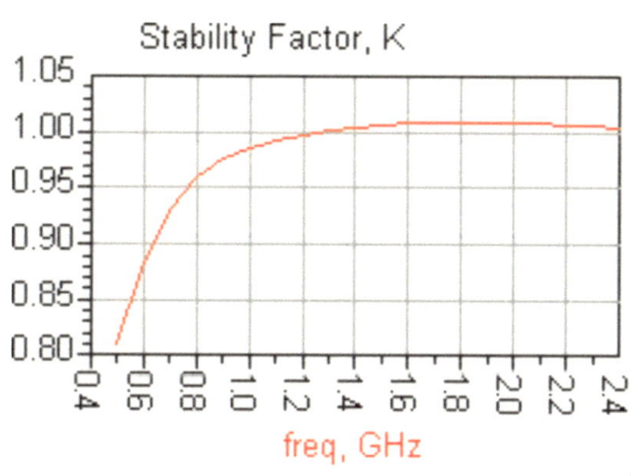

Figure 13.
ATF34143 stability results with increased source inductance.

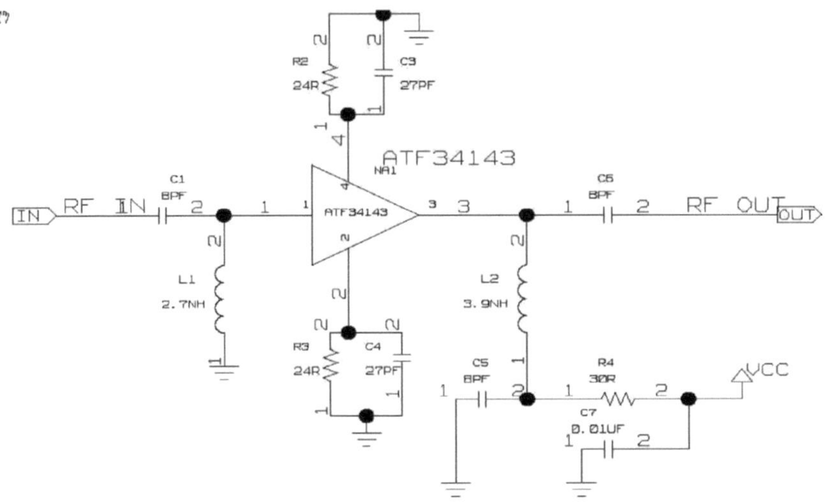

Figure 14.
Initial electrical schematic.

Figure 15.
Double-sided PCB layout (the bottom layer is a large area grounding).

3. Conclusion

It is proved that Amount and direction of velocity vectors or strain rate is proportional; to the quality and quantity of the stress impending. Shorter vectors signify of stored seismic energy at the reservoir of stratum. And hence characterizes of shallow focus expectation. An III vector of medium size signifies of outbreak stress drop at medium depth. All the vectors aligned in one direction and with same magnitude it is characteristic of stress line (tangential or radial perpendicular to the vectors observed At right angle to the vectors pattern observed at one region there lies the region of perpendicular stress viz. Tangential –to radial and vice versa. Vectors at these two region shows perpendicularity in relation. Velocity vectors of shorter size in general characterizes of foreshocks of seismicity at the region. Whereas larger size signifies greater magnitude of seismicity as in china 1996 with 7.5 Mw. Expectation of epicenter and focus lies better at the terminal point of

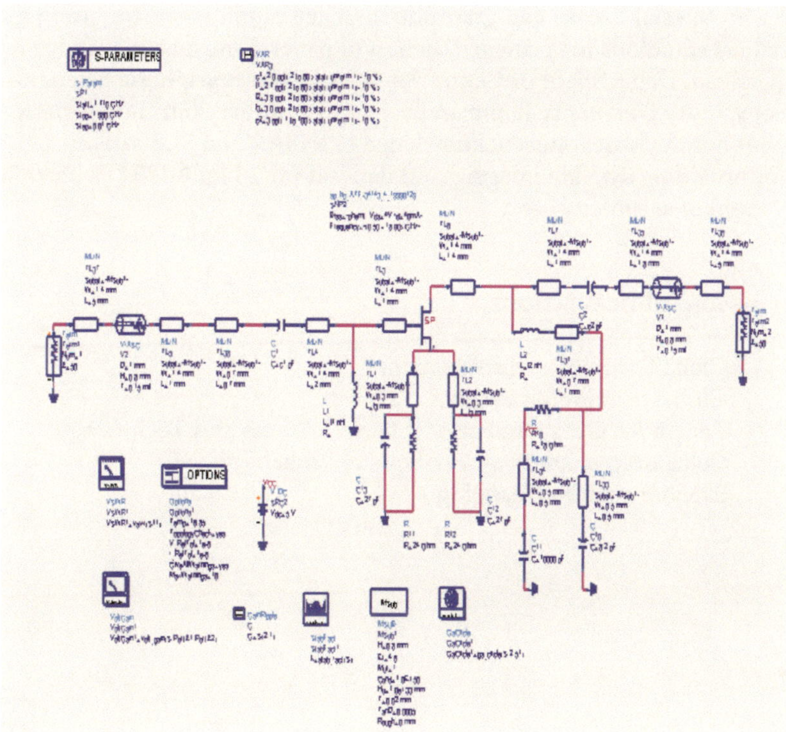

Figure 16.
Electrical schematic of PCB simulation.

smaller velocity vectors. Smaller vectors are characteristics of higher magnitude and Intensity of seismicity at the place of observation. If any motion occurs at source of waves then some surface currents are induced on some parts of the earth since the earth includes inhomogeneous materials; i.e., including conducting and/or non conducting bodies and/or free spaces, etc. in some locations. If any part of these materials makes a deviation then those surface currents variate with respect to time.

The assumption for the fact gives the result below: The Lorentz's forces, which are applied on variating currents due to the geo electromagnetic field appear at least. A new force [2] additional to Lorentz's force has to be observed due to the irregularities, too. This additional force has to have a very small magnitude around source of waves, but it has to have an irregularly and non-smoothly deviating character, so it propagates with increasing-decreasing in magnitude with some periods according to some transfer rules of forces in bodies. The modernized physical mechanism beyond the earthquake phenomena is explained with the above mentioned approach. The fact behind the earthquake phenomenon is postulated. Restrictions at the use of some specific frequency values are necessary to the frequency spectrum used in systems all around the earth to realize a successful prediction.

Acknowledgements

I give my gratitude and obligation are bid to Author cited in the reference section for their prior and later permission for theirs; reference figure and data viz.; Dr. E. Holt; Stony Brook University New York.

I offer my special thanks and gratitude to Wiley publications for citing the mathematical equations for principal section of paper. I must acknowledge to Dr. P. V.S. Raju, Head, Department of Centre for Ocean and atmospheric Sciences and Technology amity University Jaipur India, support to carry out the research.

Last but not the least I must acknowledge to INSAR and SLR satellite agency of NASA for providing satellite imagery and data set for 28 June 1992 California landers, event of seismicity.

Acronyms and abbreviations

GSRM	GlobalStrain rate Measurement
GPS	Global positioning Sysyem
COAST	Centre for Ocean and atmospheric Sciences and Technology
CSEP	Collaborative study on earthquake Probablity
SAR	Satellite acquired Ranging

Author details

Umesh Prasad Verma[1], Madhurendra Narain Sinha[2], Pushan Kumar Dutta[3]* and Subhra Mullick[4]

1 COAST Jaipur Amity University, Rajasthan, India

2 School of Engineering, Amity University, Kolkata, India

3 Geology Department, Patna University, Patna, India

4 Geological Studies Unit, Indian Statistical Institute, Kolkata, West Bengal, India

*Address all correspondence to: pkdgeoindia@gmail.com

References

[1] Allmendinger, R. W., Royse Jr, F., Anders, M. H., Christie-Blick, N., & Wills, S. (1995). Is the Sevier Desert reflection of west-central Utah a normal fault?: Comment and Reply. Geology, 23 (7), 669-670.

[2] Allmendinger, R. W., Reilinger, R., & Loveless, J. (2007). Strain and rotation rate from GPS in Tibet, Anatolia, and the Altiplano. Tectonics, 26(3).

[3] Muskhelishvili NI (1963) Some basic problems in mathematical theory of elasticity. Noordhoff International Publishing, Leyden

[4] Eringen, A. C. (1967). Mechanics of Continua. Jhon Wiley and Sons. *New York*.

[5] Timoshenko, S. and J.N. Goodier, Theory of Elasticity (McGraw- Hill New York, 1970).

[6] Thatcher, W.(1975)Strain accumulation and release mechanism of 1976 San Francisisco earthquake; J. Geophysics.Res.80,4,862-4,872.

[7] Mushkevili, N.I., Some basic Problems of Mathematical Theory of elasticity,4th Ed.(P .Noordhoff, Groningen, 1963)

[8] Morland, L. W. (1976). Elastic anisotropy of regularly jointed media.

[9] P. L. Israelevich Y. Yair, A. D. Devin, J. H. Joseph. Levin, I. Mayo.M. Molem, Transient airglow enhancements observed from the space shuttle Columbia during the MEIDEX sprite campaignGEOPHYSICAL RESEARCH LETTERS, VOL. 31, L06124, doi: 10.1029/2003GL019110, 2004.

[10] Dziewonski, A. M., Chou, T. A., & Woodhouse, J. H. (1981). Determination of earthquake source parameters from waveform data for studies of global and regional seismicity. *Journal of Geophysical Research: Solid Earth*, *86* (B4), 2825–2852.

[11] Holt, W. E., Kreemer, C., Haines, A. J., Estey, L., Meertens, C., Blewitt, G., & Lavallée, D. (2005). Project helps constrain continental dynamics and seismic hazards. Eos, Transactions American Geophysical Union, 86(41), 383-387.

[12] Zhao, J., Broms, B. B., Zhou, Y., & Choa, V. (1994). A study of the weathering of the Bukit Timah granite part A: review, field observations and geophysical survey. Bulletin of the International Association of Engineering Geology-Bulletin de l'Association Internationale de Géologie de l'Ingénieur, 49(1), 97-106.

[13] Kreemer, C., Holt, W. E., & Haines, A. J. (2003). An integrated global model of present-day plate motions and plate boundary deformation. Geophysical Journal International, 154(1), 8-34.

[14] Marrett, R., & Allmendinger, R. W. (1990). Kinematic analysis of fault-slip data. Journal of structural geology, 12 (8), 973-986.

[15] Shen Tu,B. W.E. Holt, and A.J. Haines , deformation kinetics in the western Un determined from the Quaternary slip rate and recent geodetic data .J. Geophysics,28955,1999.

[16] Price, E. and D. T. Sand well(1998): Small scale deformations associated with the 1992 Landers California earthquake mapped by synthetic aperture radar inferometry, phase gradient, J. Geophysics,res.103, 27,001-27,016.

[17] Price, E. and D. T. Sand well(1998): Small scale deformations associated with the 1992 Landers California

earthquake mapped by synthetic aperture radar inferometry, phase gradient, J. Geophysics,res.103, 27,001-27,016.

[18] . Savage .J.C. and R.O. Bur ford (1973) geodetic determination of relative plate motion in central California, J, geophysics, Res. 103, 832–845.

[19] Jaeger J.C. and N.G.W. Cook: Fundamentals of Rock Mechanics (Chapman and Hall, London,(1976)

[20] F.A. Levinzon, "Ultra-low-noise seismic piezoelectric accelerometer with integral FET amplifier," IEEE Sensor Journal, 12(6), 2012.

[21] Zechar, J. D., & Jordan, T. H. (2008). Testing alarm-based earthquake predictions. Geophysical Journal International, 172(2), 715-724.

[22] Smith, D.E.etal.(1990) ;Tectonic motion and deformation from satellite laser ranging LAGEOS, J Geophysics Res, 95,22013–22041.

[23] Pondrelli, S. A. Morelli,G. Ekstrom, S. Mazza, E .Boschi, and A.M. Dzeinweski, Eu Mediterranean regional centroid moment tensors:1997–2000, Phys, earth and planet 2002.

[24] Bieniawski, Z. T. (1967). Stability concept of brittle fracture propagation in rock. Engineering Geology, 2(3), 149-162.

[25] Kreemer, C., Haines, J., Holt, W. E., Blewitt, G., & Lavallee, D. (2000). On the determination of a global strain rate model. Earth, Planets and Space, 52(10), 765-770.

Chapter 8

Nature Inspired Metaheuristic Approach for Best Tool Work Combination for EDM Process

Goutam Kumar Bose and Pritam Pain

Abstract

As the modern day, the technologies are approaching with high accuracy at the same time with low material costing, non-traditional machining is very much essential to sustain in this modern manufacturing system. In this present research work Electric Discharge Machining (EDM) is used with different types of tools like Copper, Aluminium, and Brass are used while machining High Carbon High Chromium (HCHCr), Hot Die Steel (HDS) and Oil Hardened Nitride Steel (OHNS) workpiece material. This research work is aimed to find out the most efficient tool material for different workpiece materials while satisfying the contradictory objectives of high material removal rate (MRR) and low Tool Wear Rate (TWR). The experimental data are trained and validated by using Artificial Neural Network (ANN). Finally, the results obtained through Genetic Algorithm are hybridized with a Fuzzy- Multi Criteria Decision Making (MCDM) technique to obtain a single parametric combination of the process control parameters which satisfies these two contradictory objectives function simultaneously.

Keywords: EDM, ANN, Genetic Algorithm, Fuzzy, MDCM

1. Introduction

The progressive growth in the advanced manufacturing process is aimed to achieve a finished material with a complex shape having very high accuracy. This precise manufacturing leads the industries to modern non-traditional machining processes. Electric Discharge Machining (EDM) is one such wieldy used non-traditional machining, where material removal takes place by the control erosion through spark discharge from the cathode tool on the anode workpiece. The conductive tool and the workpiece are submerged in flowing dielectric fluid and are separated by a small gap, known as a spark gap. The temperature of the spark varies from 8000°C to 12000°C which melts and vaporizes the workpiece material instantly. This electric discharge process is used to manufacturing complex part of a metal mould, tool and die, industrial instruments, aerospace's instruments, etc. The main advantage of EDM is that this process can machine hard material accurately [1]. EDM is a complex machining process which depends upon several interrelated control parameters; hence it is important to find an optimal control parameter setting so that the machining can be optimized for a better output. In the modern era, the optimization techniques are mostly nature inspired metaheuristics process. Artificial Neural Network (ANN) is one such nature-inspired technique where the results are trained, validated and finally tested within an artificial network. In 1940, D.O. Hebb first introduced neural plasticity-based learning and then finally backpropagation is developed by Werbos in 1975 [2].

Previous research works which have been carried out to optimize the process control parameters either analytically or by simulation is presented here. Bose and Pain [3] have studied the effect of EDM on different types of tool material used in plastic industries and they have concluded different process control parameters for different material. Ho and Newman [4] have experimented on different types of an electric spark in EDM and have designed a simplified electrode to increase the performance index of the process. Zhou et al. [5] studied on minimum variance and pole placement coupled controller along with two-step prediction controllers to stabilize the machining in EDM. Equbal and Sood [6] have discussed the various parameters of the EDM process. They have also elaborated the future scope of EDM in industries. Choudhary and Jadoun [7] has experimented various types of EDM fluid in order to optimize machining productivity. They have developed die-sinking EDM, dry EDM, powder mixed EDM and also water-based EDM process. Abulais [8] has researched on various types of EDM like ultrasonic vibration dry EDM, powder-based EDM, and also used water as the dielectric fluid. Lin et al. [9] used Grey Neural Network on EDM and verified that the data are very similar to the actual experimental result. Ni [10] has discussed the various type of application of the Artificial Neural Network (ANN) in various uneasy condition and achieved a key technology for this application.

The experimental results are optimized by Genetic Algorithm (GA). Contradictory responses during machining like high Material Removal Rate (MRR) and low Tool Wear Rate (TWR) can be optimized by applying Fuzzy- Multi-Criteria Decision Making (MCDM) techniques. The objective of this research work is to identify the best tool work combination which will satisfy the contradictory responses of high MRR and low TWR.

2. Experimental design

The present research study is done on the Die Sinking EDM (Electronica make). In this research work the workpiece material, tool material, current and pulse on time (POT) are varied simultaneously. The tools considered here are Copper, Aluminum, and brass. While the workpiece material used are High Carbon High Chromium (HCHCr) steel, Hot Die Steel (HDS) and Oil Hardened Nitride Steel (OHNS). During experimentation, the current is varied in three levels as 10 amps, 15 amps and 20 amps and the Pulse on Time (POT) are varied in three levels as 800 μsec, 1600 μsec, and 2000 μsec respectively. Other parameters which have a significant effect on the process are kept constant. Here kerosene is used as the dielectric fluid, voltage is kept at 85 Volts, pulse off time is set at 800 μsec, depth of cut considered is 2 mm and spark gap is 5 mm. During the experimental run, various parameters are varied simultaneously by following the L9 Orthogonal Array (OA) so that the experimental time and as well as experimental cost can be reduced to a great extent. To analyze the data statistically the tool materials and also workpiece materials are expressed by their respective density as in **Table 1**.

From the experimental run the regression equitation is obtained for the MRR, where a1, b1, c1 and d1 are constant terms:

$$\begin{aligned} \text{MRR} = {} & \text{a1} + \text{b1} * \text{Tool Density} + \text{c1} * \text{W/P Density} + \text{d1} * \text{Current} + \text{e1} * \text{POT} \\ & + \text{f1} * \text{Machining Time} \end{aligned} \tag{1}$$

Materials	Symbols	Density (g/cm^3)
Tool Materials	Cu	8.96
	Al	2.78
	Br	8.73
Work Materials	HCHCr	7.7
	HDS	7.75
	OHNS	8.67

Table 1.
Density of the tool and workpiece materials.

From the same experimental run Tool Wear Rate (TWR) is obtained where a2, b2, c2 and d2 are constant terms:

$$\text{TWR} = \text{a2} + \text{b2} * \text{Tool Density} + \text{c2} * \text{W/P Density} + \text{d2} * \text{Current} + \text{e2} * \text{POT} + \text{f2} * \text{Machining Time} \tag{2}$$

Here Artificial Neural Network (ANN) is utilized to test and validate the experimental data. Then the responses Material Removal Rate (MRR) and Tool Wear Rate (TWR) are optimized by Genetic Algorithm (GA) in 'MATLAB R2015a' environment. The global optimal solution is then calculated by the Multi-Criteria Decision Making (MCDM) technique by applying Fuzzy set theory. Finally, the calculated data are analyzed by actual experimentation to validate the final result.

3. Artificial Neural Network

Artificial Neural Network (ANN) is the replica of the actual neural network system and there working principle is quite similar to the biological neural network [11]. Outer nodes collect the input responses. The other nodes are inter-connected, and finally, nodes give the responses to the input responses. In a neuron, there are synapses, they multiply each input by the weighted value, only if this value receded the threshold value, then these responses transfer to the next neuron. The interconnected inner layer of the neuron is known as a hidden layer. Eq. (3) shows the input calculation for each hidden layer of the neuron.

$$\boldsymbol{II_i} = \sum_{i-1}^{n} \boldsymbol{W_i X_i} \tag{3}$$

The output responses are defined by sigmoid function as shown in Eq. (4).

$$\boldsymbol{O_i} = \boldsymbol{f(I_i)} = \frac{\mathbf{1}}{\mathbf{1} + \boldsymbol{e^{-I_i}}} \tag{4}$$

The working principle of an artificial neuron is shown in **Figure 1**.
There are generally three types of architectures in case of Neural Network.

- Feedforward-neural networks
- Feedback-neural networks

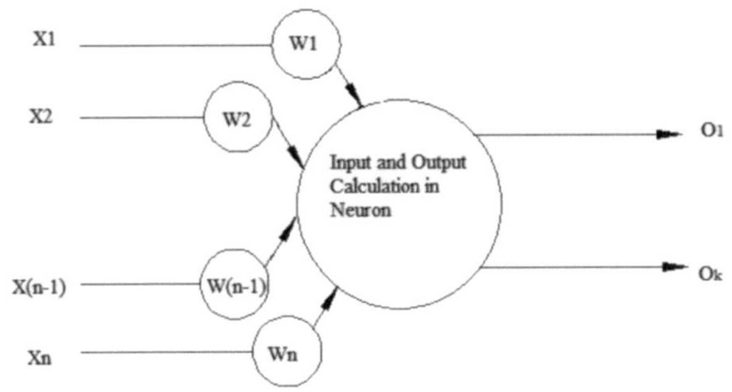

Figure 1.
Schematic diagram of an artificial neural.

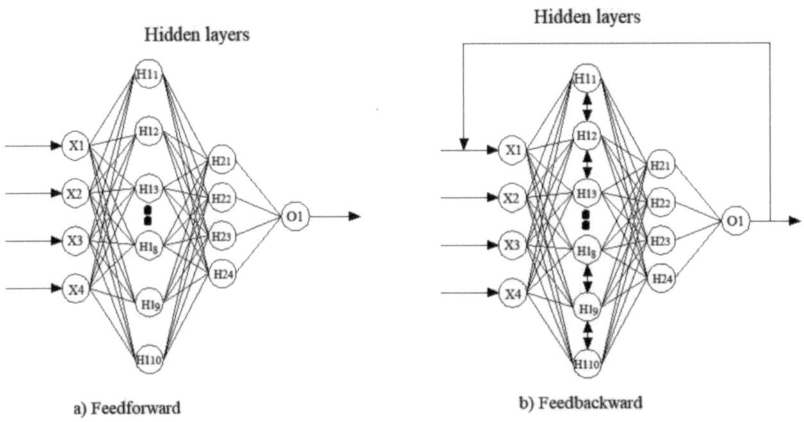

Figure 2.
The architecture of general ANN.

- Self-organizing-neural networks

Figure 2 displays the general structure of ANN.

3.1 Analysis of the experimental result

In this research work, the analysis of the experimental results is analyzed by dividerand function, in using Matlab R2015a. The backpropagation method is used in this analysis because this backpropagation method gives the feedback while training the data for calculation. The individual responses are evaluated using the Simulink model in ANN considering all those five parameters. Also, ten number of hidden layers are used to optimize the responses. **Figure 3** shows the Simulink diagram used in ANN.

In ANN responses are trained then validated and finally tested to find out that is there any linear relationship between control parameters and responses. The continuous line is the best fit linear regression line of output versus targets. While the value of regression (R) is 1 represents the linear relation between control and response parameters.

3.1.1 Analysis to maximize MRR

For calculating 50%, 25% and 25% of the date have been used for training, testing and validation, respectively. MSE has been achieved after 5 successful

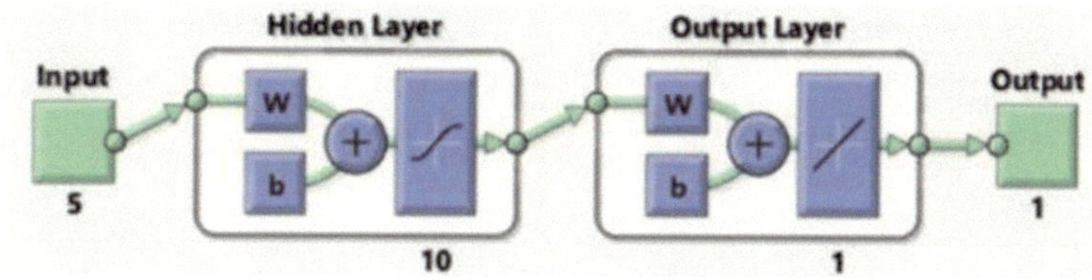

Figure 3.
Simulink diagram of ANN.

iteration and as a result the training has been terminated. It has been programmed in such a way that in case the gradient falls below 1.00×10^{-7}, the training will terminate. In this experiment the gradient is 5.01×10^{-8}. Here the number of successive iterations performed for validation checks is 2. **Figure 4** represents the neural network training performance progress.

Best validation performance is 5.0748×10^{-7} at epoch 3, which is a low prediction error measured with MSE shown in **Figure 5**. This graph does not show any main problems with the training. The validation and test curves are very similar to each other up to 3 epochs.

In this case, the values of R for training, testing and validation are 0.85099, 1 and 1, respectively. So, from the **Figure 6** it is evident that the value of overall R is 0.77755. As the values of R for validation and testing both has the values more than 0.9 the training shows a good result.

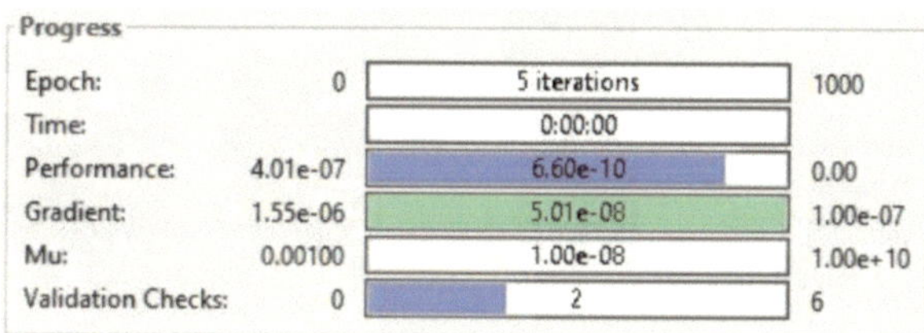

Figure 4.
Training performance progress for MRR.

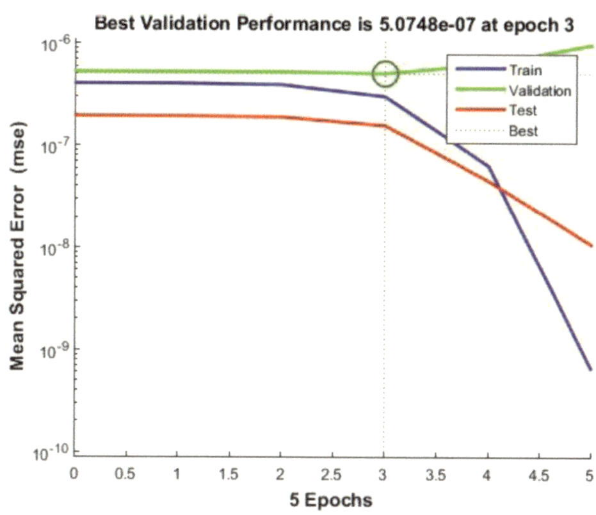

Figure 5.
Performance plot for MRR.

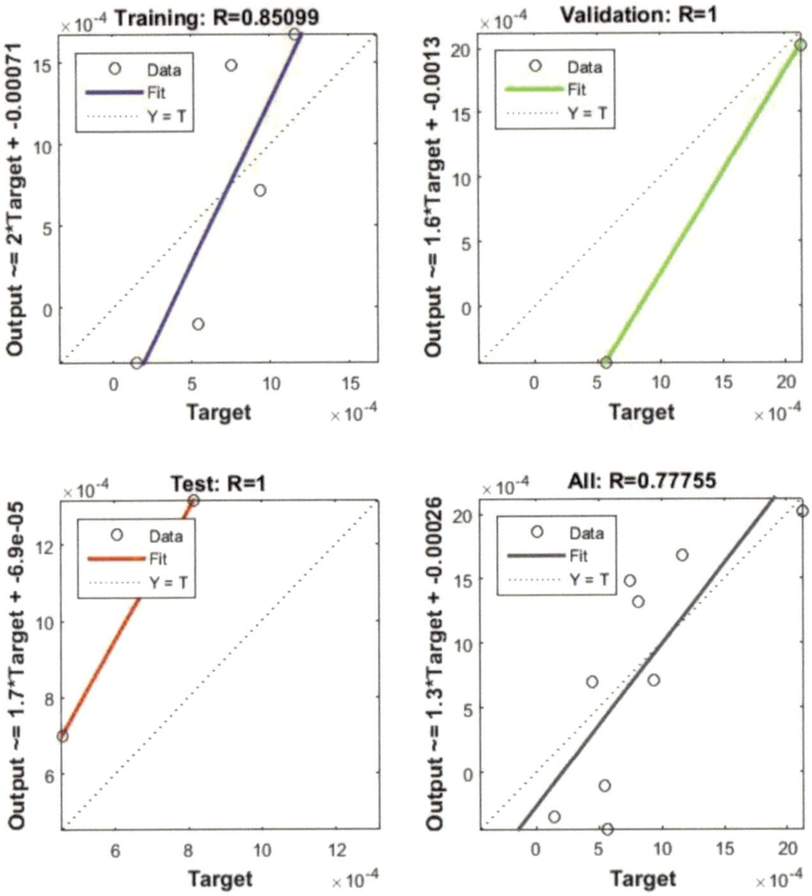

Figure 6.
Regression plot of MRR.

3.1.2 Analysis to minimize TWR

For calculating 55%, 25% and 20% of the date have been used for training, testing and validation, respectively. MSE has been achieved after 4 successful iteration and as a result the training has been terminated. It has been programmed in such a way that in case the gradient falls below $1.00 \text{x} 10^{-7}$, the training will terminate. In this experiment the gradient is $7.63 \text{x} 10^{-8}$. **Figure 7** represents the neural network training performance progress.

Best validation performance is $206859 \text{x} 10^{-5}$ at epoch 4, which is a low prediction error measured with MSE shown in **Figure 8**. This graph does not show any main problems with the training. The validation and test curves are very similar to each other.

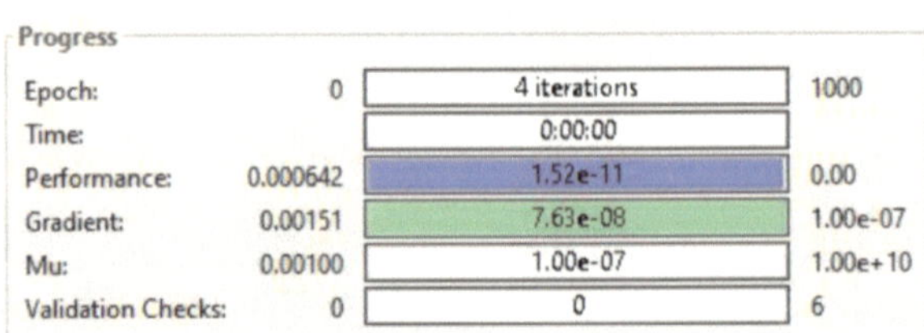

Figure 7.
Training performance progress for TWR.

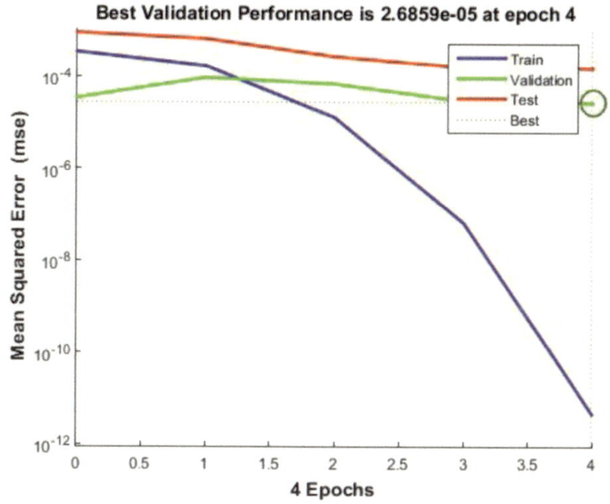

Figure 8.
Performance plot for TWR.

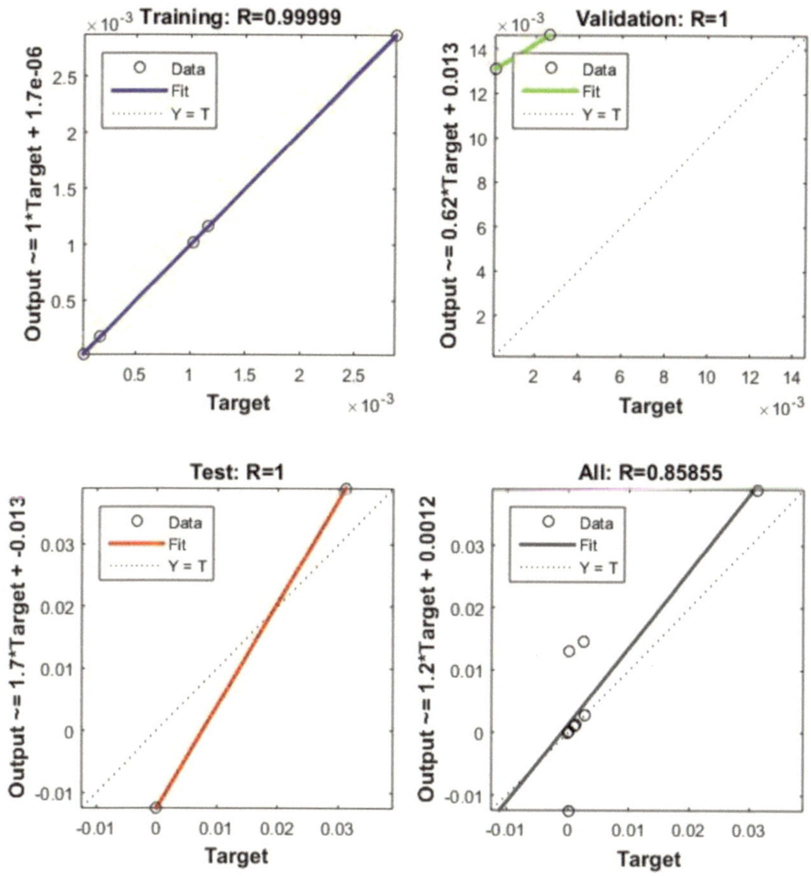

Figure 9.
Regression plot of TWR.

In this case, the values of R for training, testing and validation are 0.99999, 1 and 1, respectively. From the **Figure 9**, the value of overall R is 0.85855. As the values of

R for validation and testing both has the values more than 0.9 the training shows a good result.

4. Genetic Algorithm (GA)

Genetic Algorithm is a similar approximation method as survival of the fittest. This nature-inspired metaheuristic process follows some fundamental rules [12].

- Every living being in any ecosystem struggle for food and mates.
- Those victorious individuals will compete and fittest offspring will be produced. Those individuals also are known as 'King' in the system.
- A good individual will reproduce again and again and will survive in nature for a long time than those week individuals.

Each fitness functions are considered as individual chromosome and they are the various sequence of binary alphabets (1 and 0). This algorithm is very useful to find out the global solution to a problem set.

On average, better new generations are formed with better genes. Every successive generation will have a 'partial better solution' than the previous generations. Ultimately, when the newly created offspring does not have a noticeable difference from the previous generations, then the algorithm will terminate at a converged solution.

4.1 Results using the Genetic algorithm

It has been aimed to find out the single parametric combination for this contradictory parameters by using multi-response optimization. In order to find out the two contradictory parameters like high MRR and low TWR. The boundary condition for this genetic algorithm is used as follows:

- Population

  Population type: Double vector

- Stopping Criteria

  Generations: 100 x number of variables

  Time limit: infinity

  Stall generations: 100

  Function tolerance: $1\text{x}10^{-4}$

  Constrain tolerance: $1\text{x}10^{-3}$

Here the total number of iterations required for optimization is 127 and which gives 16 combinations for the control parameters along with responses. Optimization terminated as the average change in the spread of Pareto solutions has been reached to its tolerance value. Based on the conflicting nature of the objectives, multi-response optimization is carried out in order to achieve the goal by a single parametric combination. As the EDM process is complex machining, it can have two general situations while it is used for commercial purposes. In one case the

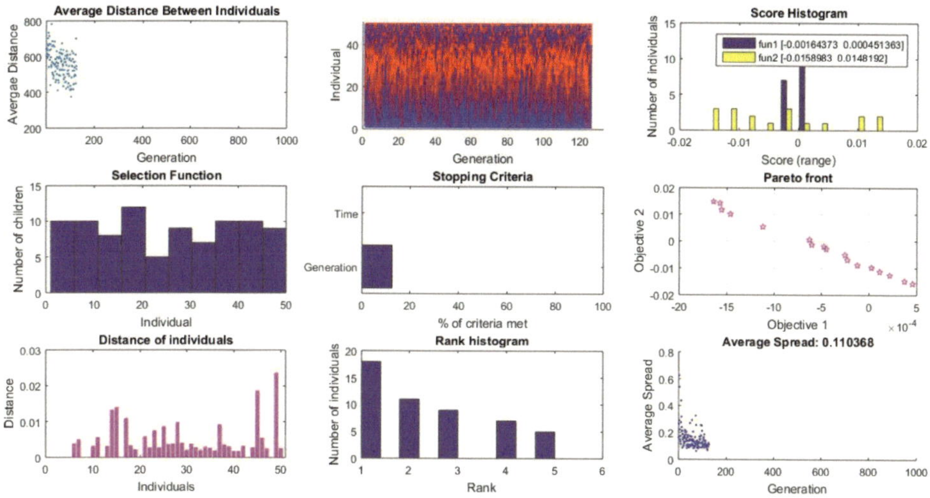

Figure 10.
Plot functions for GA.

primary objective is to achieve maximum MRR. It can be used for rough cutting. In another case for finishing operation, more emphasis should be given on TWR instead of MRR. In this condition, the tool wear can highly affect the final geometry of the product. As the genetic algorithm is generally subjected to minimizing the function, so in the case of maximizing the MRR the negative sign have been neglected. In **Figure 10** the two contradictory objectives are simultaneously optimized by using GA have been plotted. From this plot, the boundary condition for

Exp. No	Tool Density (g/cm^3)	W/P Density (g/cm^3)	Current (Amp)	POT (μSec)	Machining Time (min)	MRR (cm^3/min)	TWR (gm/min)
1	4.5	15.90	88.58	864.64	101.16	1045.33	1216.031
3	16.4	14.82	89.36	771.19	198.54	1436.91	149.4018
4	0.2	9.78	87.15	859.08	126.59	1100.34	961.2881
5	15.7	14.17	86.24	783.69	199.16	1437.78	149.4643
6	15.6	11.71	88.01	774.51	194.3	1167.08	185.9447
7	2.2	7.04	86.72	826.90	158.6	1074.93	1057.663
8	4.5	2.85	88.46	819.28	151.4	1201.48	827.6156
9	2.5	5.09	88.50	828.50	102.9	1145.31	683.333
10	6.3	0.67	88.23	800.79	119.1	1233.71	531.6255
11	2.1	12.68	88.31	849.67	123.3	1070.92	1178.077
12	6.0	1.13	88.41	828.16	150.9	1154.8	633.3167
13	4.7	1.85	88.44	814.21	158.9	1305.47	875.2726
14	1.1	11.47	88.40	860.27	129.6	1072.24	1070.531
15	3.7	14.69	88.50	863.51	101.2	1049.99	1138.689
16	11.2	5.36	88.51	786.64	175.8	1111.97	461.5821
17	1.2	8.81	88.59	851.90	158.6	1074.93	1057.913

Table 2.
Combination of factors and responses.

MRR and TWR can be found out. For MRR the range is between 1045.3295 cm^3/min to 1437.789 cm^3/min and for TWR, it varies between 149.402 gm/min to 1216.031 gm/min respectively. Therefore, in order to arrive at an optimal or near optimal parametric combination which will consecutively satisfy the contradictory nature of the responses Fuzzy Gray Relational Analysis is conducted.

The different parametric combinations with respective responses as obtained through GA are shown in **Table 2** below.

5. Multi objective solution using Grey Relation Analysis (GRA)

GRA can be employed to simultaneously find out the optimized solution for several contradictory responses [13]. This theory has been proposed by Deng in 1982's [14]. In modern research work, this theory is a very essential tool to design the model for the unknown of partially known or unspecified data.

GRA form the link between preferred (best/ideal) with real investigational data. The average of the grey coefficient is used to estimate the grey grade. This grade is generally varies between 0 and 1. When the value is close to 1, it signifies that the solution approaches the ideal condition. In the final acquired data set the parametric combinations which have the maximum grey relation grade, that combination will be termed as the optimized solution.

The normalized equation for the condition where the maximum value is required, like MRR that can be expressed as:

$$X_{ij} = \frac{Y_{ij} - Min\left[Y_{ij}, i = 1, 2, \ldots \ldots n\right]}{Max\left[Y_{ij}, i = 1, 2, \ldots \ldots n\right] - Min\left[Y_{ij}, i = 1, 2, \ldots n\right]} \tag{5}$$

If lower value for the better performance such as TWR then it is expressed as,

$$X_{ij} = \frac{Max\left[Y_{ij}, i = 1, 2, \ldots \ldots n\right] - Y_{ij}}{Max\left[Y_{ij}, i = 1, 2, \ldots \ldots n\right] - Min\left[Y_{ij}, i = 1, 2, \ldots n\right]} \tag{6}$$

To find out the single solution for these two contradictory processes the GRA is performed. When the grey grade is 1, that solution gives the optimized single parametric combination.

6. Fuzzy set theory

To find out the ambiguous solution in decision-making problem, the Fuzzy set theory can be used as a powerful tool. Instead of using numerical values, assign of weights for linguistic assessment is more useful [15]. During the consideration of the decision makers' fuzzy rating, fuzzy decision matrix can be achieved from a decision matrix and finally it can be converted into weighted normalized fuzzy decision matrix. A fuzzy set can be described by a membership function $\mu_{\hat{d}}(x)$ while converting X. A degree of membership of x in $\hat{d}$ can be plots individual element x in X to a real number in the period of 0 to 1. In this case triangular fuzzy number (TFN), can be defined as a triplet (d1, d2, … , dn) and the membership function is defined [16].

The translation method of fuzzy number into the non-fuzzy number, that is, a crisp value is identified as defuzzification. In this current research work 'centroid of area' technique for defining Best Non-Fuzzy Performance (BNP) value is applied.

7. Multi-criteria decision making (MCDM) analysis

In this chapter the two contradictory responses i.e. MRR and TWR have got a dissimilar level of rank. In this case the maximized MRR is the primary objective rather than the lower TWR.

Table 3 is tabulated by using the weightage from the fuzzy set theory.

The closed value to the 1 gives the ideal solution between the comparative sequences. In this case, 16 simulated data from the genetic algorithm have been used for further evaluation. In this case, the criteria for decision making have been set as Maximum MRR and minimum TWR.

7.1 Optimization of the parameters

The specified weights for MRR and TWR are 84.4% and 15.6% respectively as calculated by using Entropy method [17]. **Table 4** represents the grey relation coefficient and grades corresponding to parametric settings and responses for the material in **Table 2**.

Criteria	Linguistic terms	Fuzzy number	BNP
MRR	VH	0.9,1.0,1.0	0.844
TWR	ML	0.1,0.3,0.5	0.156

Table 3.
Weight criteria for deference responses.

Exp. No	Responses		Grey Co-efficient		Grey Grade	Rank
	MRR (cm^3/min)	TWR (gm/min)	MRR (cm^3/min)	TWR (gm/min)		
1	1045.330	1216.031	0.458	0.135	0.296	16
2	1436.914	149.402	0.997	1.000	0.999	2
3	1100.345	961.288	0.495	0.170	0.333	10
4	1437.789	149.464	1.000	1.000	1.000	1
5	1167.082	185.945	0.550	0.820	0.685	3
6	1074.937	1057.663	0.477	0.155	0.316	11
7	1201.488	827.616	0.584	0.197	0.390	8
8	1145.313	683.333	0.531	0.238	0.384	9
9	1233.717	531.626	0.619	0.303	0.461	4
10	1070.928	1178.077	0.474	0.139	0.307	14
11	1154.800	633.317	0.539	0.256	0.398	7
12	1305.472	875.273	0.715	0.186	0.451	5
13	1072.248	1070.531	0.475	0.153	0.314	13
14	1049.991	1138.689	0.461	0.144	0.302	15
15	1111.972	461.582	0.504	0.348	0.426	6
16	1074.937	1057.913	0.477	0.155	0.316	12

Table 4.
Grey relation co-efficient along with grades and ranks.

Settings Levels	Predicted result	Experimental result
MRR	1437.789	1385.74
TWR	149.464	256.84
Grey Grade	1.000	0.736
Improvement of the grey relation grade: 0.264		

Table 5.
Results of machining performance using the initial and optimal machining parameters.

Grey relation coefficient, relation grade, and the ranks have been displayed in **Table 4**. From this table, it is obvious that the experimental run number 4 has achieved maximum gray relation grade. As has been discuses earlier, this experimental run satisfies the condition for the optimized multi-response parameter. So, experiment 4, which have parametric combination Tool Density 15.7 g/cm^3, Workpiece density of 14.17 g/cm^3, Current 86.24 amp, POT 7836.89 μSec, and Machining time of 199.16 min is the best parametric combination for having high MRR and low TWR.

The confirmation experiment performed with the above optimal combination results in grey relational grade MRR and TWR is obtained as 1385.75 cm^3/min and 256.84 gm/min respectively. It is observed that MRR and TWR improve significantly by using optimal machining variables combinations. **Table 5** shows the validation results while machining at optimizing condition.

8. Conclusion

The experimental study indicates that while machining different workpieces like HCHCr, HDS, and OHNS using different EDM tools like Cu, Al, and Br, the responses are dependent on tool material, workpiece material, pulse on time, and machining time. While analyzing the response data individually applying, ANN considering four control parameters in order to achieve maximum MRR and minimum TWR the training, validation, and testing data indicates that the values of R, are almost 1.

A multi-objective response is developed which is optimized using GA. Since the objectives of the responses are contradictory in nature, therefore, using GA the optimum values of responses are obtained within a range. In the case of MRR varies from 1045.3295 cm^3/min to 1437.789 cm^3/min and TWR varies between 149.402 gm/min to 1216.031 gm/min respectively.

The GRA establishes the ranks of output for different variables combinations and establishes optimal combinations for a complex process like EDM process. For evaluating the optimum parametric combination during machining her Tool Density of 15.7 g/cm3, Workpiece density of 14.17 g/cm^3, Current of 86.24 amp, POT of 7836.89 μSec and Machining time of 199.16 min is the best among all the other combinations for having high MRR and low TWR. If any tool material and workpiece have the exact tool density as obtained from the GA that will give the best tool and workpiece combination for machining. As our research is limited to the three types of tool material and three types of the workpiece material, hence it can be concluded that Cu is the best tool for machining OHNS workpiece material by EDM when 20 amp current and POT is 800 μSec.

Therefore, this experimental analysis for estimating the optimum EDM parametric combination during machining with a different tool and work materials can act as valuable and an effective guideline for machining of die and mould.

Author details

Goutam Kumar Bose and Pritam Pain*
Department of Mechanical Engineering, Haldia Institute of Technology, Haldia, West Bengal, India

*Address all correspondence to: pritam.me.dscsdec@gmail.com

References

[1] El-Hofy, H.: Advanced Machining Processes: Nontraditional and Hybrid Machining Processes, Mcgraw-hill, (2005).

[2] Werbos, P. J.: Generalization of backpropagation with application to a recurrent gas market model, Neural Networks, Vol. 1(4), (1988) 339–356.

[3] Bose. G.K. and Pain. P.: Parametric Analysis of Different Grades of Steel Materials Used in Plastic Industries through Die Sinking EDM Process. International Journal of Materials Forming and Machining Processes, Vol. 3(1), (2016) 55–74.

[4] Ho, K. H. and Newman, S. T.: State of the electrical discharge machining (EDM), International Journal of Machine Tools and Manufacture, Vol. 43(13), (2003) 1287–1300.

[5] Zhou, M., Wu, J., Yang, J. and Yao, D.: Fast and Stable Electrical Discharge Machining (EDM) by Two-step-ahead Predicted Control, Procedia CIPR, Vol. 42, (2016) 215–220.

[6] Equbal, A. and Sood, A.K.: Electrical Discharge Machining: An Overview on Various Areas of Research, Journal of Manufacturing and Industrial Engineering (MIE), Vol. 1-2(13), (2014) 1–6.

[7] Choudhary, S. K. and Dr. Jadoun, R.S.: Current Advanced Research Development of Electric Discharge Machining (EDM): A Review, International Journal of Research in Advent Technology, Vol. 2(3), (2014) 271–297.

[8] Abulais, S. Current Research trends in Electric Discharge Machining (EDM): Review. International Journal of Scientific & Engineering Research, Vol. 5(6), (2014) 100–118.

[9] Lin, Z., Liu, Y. and Zhang, L.: Research of EDM (Electrical Discharge Machining) Process Simulation Based on Grey Neural Network. Advances in Mechanical and Electronic Engineering. Lecture Notes in Electrical Engineering, Vol. 177, (2012) 373–379.

[10] Ni, X.: Research of Data Mining Based on Neural Networks, World Academy of Science, Engineering and Technology, Vol. 39, (2008) 381–384.

[11] McCulloch, W., Walter, P.: A Logical Calculus of Ideas Immanent in Nervous Activity. Bulletin of Mathematical Biophysics. Vol. 5 (4). (1943) 115–133.

[12] Goldberg, D. E.: Genetic Algorithms in Search, Optimization and Machine Learning. Addison-Wesley Longman Publishing Co., Inc. Boston, MA, USA, 1989.

[13] Bose, G. K. and Pain, P.: Nature-Inspired Metaheuristic Approach for Multi-Objective Optimization During WEDM Processes. In: Ram, M. and Davim, J. P. editors. Soft Computing Techniques and Applications in Mechanical Engineering. IGI Global, 2018, pp. 91–122. DOI: 10.4018/978-1-5225-3035-0.ch004

[14] Deng, J.: Control problems of grey systems. Systems and Control Letters, 1 (5), (1982) 288–294.

[15] Kaufmann, A. and Gupta, M. M.: Introduction to Fuzzy Arithmetic: Theory and Applications, Van Nostrand Reinhold Company, 1985.

[16] Dubois, D. and Prade, H.: Operations in a fuzzy-valued logic. Information and Control, Vol 43(2), (1979) 224–240.

[17] Zhuang, X., Zhao, Y. & Huang, T. S.: A neural net algorithm for multidimensional maximum entropy spectrum estimation, Neural Networks, Vol. 4, (1991) 619–624.

Chapter 9

Atomistic Mathematical Theory for Metaheuristic Structures of Global Optimization Algorithms in Evolutionary Machine Learning for Power Systems

Jonah Lissner

Abstract

Global Optimization in the 4D nonlinear landscape generates kinds and types of particles, waves and extremals of power sets and singletons. In this chapter these are demonstrated for ranges of optimal problem-solving solution algorithms. Here, *onts*, particles, or atoms, of the ontological blueprint are generated inherently from the fractional optimization algorithms in Metaheuristic structures of computational evolutionary development. These stigmergetics are applicable to incremental machine learning regimes for computational power generation and relay, and information management systems.

Keywords: metaheuristic, particle swarm optimization, global optimization, machine learning, evolutionary algorithms, stigmergetics, arrow's paradox, atomistics, Fermat's last theorem, network theory, Hensellian mathematics

1. Introduction

The evolution of Algorithms from a simple route, to complexified paths requires maps from zones of optimal utilization, to be solved sufficiently, in a given amount of time.

These algorithms are constructed for the purpose of building and advancing a continuity for the next location of optimal utilization, in order to realize the importance to form workable nodes and circuits, that are discrete and exact algorithm criteria in a time-basis. Therefore, a complete network on a nonlinear surface and related machine learning epochs is built.

These criteria are based on Fermat's Theorem proving global extrema locations either at stationary or bounding points, based ultimately upon the Pythagorean Theorem, where:

Let **N** be the set of natural numbers 1, 2, 3, …, let **Z** be the set of integers 0, ± 1, ± 2, …, and let **Q** be the set of rational numbers a/b, where a and b are in **Z** with $b \neq 0$. In what follows we will call a solution to $x^n + y^n = z^n$ where one or more of x, y, or z is zero a *trivial solution*. A solution where all three are non-zero will be called a *non-trivial* solution [1].

IntechOpen

2. Materials and methods

2.1 Metaheuristic structural rules for the algorithm building

It is a rule of No Unreasonable Effectiveness of Mathematics in any Science [Wigner] [2], and therefore a notion that No Unreasonable Effectiveness of Axiomation in any Science, that 3 Rules of Information Physics [3IP] by Jonah Lissner exist:

I. Problem of Demarcation.

II. Rule of Information Dichotomy [*Gestalt-Inverse Gestalt*], and thereby requiring a kind of.

III. Context-Restricted Deep Structure [CRDS] for the given topology.

Therefore hypothesized to be commutable terms within this 3IP rulebase, The Three Thermodynamic Rules of Macrodynamics by Jonah Lissner. The 3 General Rules of Macrodynamics [3GRM] which are established to define Global Optimization Algorithm [GOA] challenges:

1. The Rule of the Continuity of Primaries

2. The Rule of Perpetuation of Information Inequalities of Primaries

3. The Rule of Unprovable Ideals, Cardinals or Delimitations of Complex Adaptive Evolutionary Systems [CAES].

3. Algorithm definition

The **Metaheuristic Algorithm** is defined by the Author [Jonah Lissner] as.

An ontological mechanism to generate or activate decision paths [algorithms] and make decision potential to solve [essentially two-state] paradoxes, a computational physical network and topos, for practical effort or application. Therefore can be constructed a theoretical or hypothetical guideway from objects and particles to advance ontological gradations of relevance and value, through a logical progression.

A relevant algorithm to solve for discrete stigmergetics in nonlinear optimization challenges for graphing algorithms of power systems has been demonstrated in Ant Colony Optimization [ACO]:

Here in general formula where

$$p_{xy}^{k} = \frac{\left(\tau_{xy}^{\alpha}\right)\left(\eta_{xy}^{\beta}\right)}{\sum_{z\in \mathrm{allowed}_x}\left(\tau_{xz}^{\alpha}\right)\left(\eta_{xz}^{\beta}\right)}$$

by trail update action $\tau_{xy} \leftarrow (1-\rho)\tau_{xy} + \sum_k \Delta\tau_{xy}^{k}$.

for given function sets

$$f(x) = \lambda x, \quad \text{for } \mathrm{x} \geq 0; \tag{1}$$

$$f(x) = \lambda x^2, \quad \text{for } \mathrm{x} \geq 0; \tag{2}$$

$$f(x) = \begin{cases} \sin\left(\frac{\pi x}{2\lambda}\right), & \text{for } 0 \le x \le \lambda; \\ 0, & \text{else} \end{cases} \tag{3}$$

$$f(x) = \begin{cases} \pi x \sin\left(\frac{\pi x}{2\lambda}\right), & \text{for } 0 \le x \le \lambda; \\ 0, & \text{else} \end{cases}. \tag{4}$$

Evolutionary Game Theory [EGT] challenges in optimization schedules are therein linked to Ant Colony Optimization [ACO], e.g. Eigenvector centrality formula

$$\textbf{Where for } \mathbf{G := (E, V) \text{ with } V \text{ vertices let } A = (av, t), e.g. av, t = 1 \text{ or } av, t = 0. \text{Therefore } xv = 1/l} \\ \mathbf{SigtsetM(v)xt = 1/l \text{ SigtsetG } av, t\ xt}[ibid]. \tag{5}$$

It is a basis for optimization schedules that there is an asymmetrical velocity, mass and gravity of said scope of systems. At various times in the computational history, particle optimization on the manifolds evolve at a faster rate [or slower rate] than before. Hence the given incremental and discrete rate of increase, in valleys and peaks accelerates and stabilizes at a higher positive, null or negative value and result in extremal mechanics and nonlinear dynamics. An example can be demonstrated utilizing power faults and extremals on the electrical circuits [3].

These problems of prediction for probability of choice of one object or particle of a set, for pariwise sets and in algorithms, have been demonstrated in Arrow's Impossibility Theorem and for Algorithmic Information Theory [AIT] whence we can replace *voter* for global optimization *particle* and replace *group* with *set*:

1. If every voter prefers alternative X over alternative Y, then the group prefers X over Y.

2. If every voter's preference between X and Y remains unchanged, then the group's preference between X and Y will also remain unchanged (even if voters' preferences between other pairs like X and Z, Y and Z, or Z and W change).

3. There is no "dictator": no single voter possesses the power to always determine the group's preference [4].

Important is Criteria 3, from whence adaptive and efficient algorithms have space to be constructed as particles on the run-time, for a given Global Optimization Algorithm [GOA].

4. Building the algorithm

In a praexological theory [5] this is proposed because of the inherent general inaccuracy of specific problems, learning rubrics, and Macrodynamic properties of a given performance landscape, and ultimately inefficient of any algorithmic system, given isomorphic [atomistic or non-atomistic] qualities of rulebase, algorithmic structure, weights, and variables [6]. These in turn can be represented as information sets, materiel, work, and symbolic representation and/or power in specific *qualia* of *Historical Rule of Perpetuation of Information Inequalities* set to various scales and models.

Clerc has demonstrated a general Metaheuristic algorithm where for $f\colon \mathbb{R}^n \rightarrow \mathbb{R}$ essentially $f(\mathbf{a}) \leq f(\mathbf{b})$. S includes the number of particles in the swarm having specific position and velocity in the search—space:

```
for each particle i = 1, ..., S do
   Initialize the particle's position with a uniformly distributed random vector: xi
   ~ U(blo, bup)
   Initialize the particle's best known position to its initial position: pi ← xi
   if f(pi) < f(g) then
      update the swarm's best known position: g ← pi
   Initialize the particle's velocity: vi ~ U(−|bup-blo|, |bup-blo|)
while a termination criterion is not met do:
   for each particle i = 1, ..., S do
      for each dimension d = 1, ..., n do
         Pick random numbers: rp, rg ~ U(0,1)
         Update the particle's velocity: vi,d ← ω vi,d + φp rp (pi,d-xi,d) + φg rg
         (gd-xi,d)
      Update the particle's position: xi ← xi + vi
      if f(xi) < f(pi) then
         Update the particle's best known position: pi ← xi
         if f(pi) < f(g) then
            Update the swarm's best known position:
```

$$g \leftarrow pi.[7] \tag{6}$$

5. Results and discussion

5.1 Complex adaptive evolutionary system: thermodynamic landscape

For the purpose of evaluating algorithmic fitness on the given landscape the following ansatz can be utilized for Particle Swarm Optimization [PSO] for a decision tree:

$$\textbf{For} f : \textbf{Mn} > > \textbf{M and where}'\textbf{M} = \textbf{KH\_(t,h) x iE} \tag{7}$$

Where
M = Manifold
'M = Algorithmic Landscape/Manifold, for
KH = potentiated Knowledge History of the Algorithm
t = tradition or procedural process structure [word, string, grammar, memory mapping, rulebase, database] of the algorithm
h = computational multiplicatives and inequalities of the x iE = Adaptive Landscape History of the [Numerical] Object(s) or Particle(s) on the Network.

Lovbjerg and Krink demonstrate for the thermodynamic variables on the Particle Swarm Optimization [PSO]:

$$\rightarrow \text{vi} = \chi(\text{w} \rightarrow \text{vi} + \varphi \rightarrow 1\text{i}(\text{p} \rightarrow \text{i}- \rightarrow \text{xi}) + \varphi \rightarrow 2\text{i}(\text{p} \rightarrow \text{g}- \rightarrow \text{xi})) \tag{8}$$

where χ is the constriction coefficient.

Here the Three General Rules of Macrodynamics [3GRM] Macrodynamic Automata Rules [MAR] are applied to Information Natural Dynamics [IND] criteria:

Natural Sets, Natural Kinds, Natural Procedures, Natural Strings, Natural Radicals, Natural Binaries, Natural Radices, in Complex Adaptive Evolutionary System [CAES]-Multiagent System [MAS] for 4D model variables.

These variables should each contain criteria:

Time-Complexity, Particle-Value, Particle-Weighting in Fuzzy set theory, Gravity of System, <<> > Nanodynamics of System variables [TC-PV-PW-GS-NDS]

Set approximate to

$$f : \textbf{Omega set Rn} >> \textbf{R with the global minima} f^* \textbf{ and the set of all global minimizers } X^* \textbf{ in Omega to find the minimum best set in the function series of} (\mathbf{x}) \tag{9}$$

for system conditions, system boundaries, number and density of particles in the total Information Natural Dynamics [IND] of the Global Optimization Algorithm [GOA]. These are applied to algorithmic manifold for the candidate solution on the given search spaces. It can be argued that given the extremes of information disequilibrium applied to macrodynamic disequilibrium models, there are inevitably generated extremals of various degrees of power, in the incremental Information Dynamics.

5.2 Complex adaptive evolutionary system: weighting

These differentiable functions can be further defined c.f. Dense heterarchy in Complex Systems Algorithms of a coupled oscillators, where in general formula.

$$\frac{dx}{dt} = (P(t) + \mu\, Q(t, \mu))x + f(t), \tag{10}$$

Here in a differential equation we can demonstrate.

$$u^{(n)} = f\left(t, u, u', \dots, u^{(n-1)}\right), n \geq 2, \tag{11}$$

These can be demonstrated in Particle Swarm Optimization [PSO], and Macrodynamic models of Meta-optimization of Particle Swarm Optimization [PSO] [7], c.f.

$$\mathbf{v}_i(t+1) = w \cdot \mathbf{v}_i(t) + n_1.r_1.\left(p_i - x_i(t)\right) + n_2.r_2.\left(p_{best} - x_i(t)\right) \tag{12}$$

for each set of given epoch or evolutionary landscape scenario prediction in analytical and expectation weighting parameter formula algorithm optimization [Meissner, et al., *ibid*].

5.3 Complex adaptive evolutionary system: thermodynamics

Regarding bounding definitions, Chaitin demonstrated in Algorithmic Information Theory [AIT] algorithmic decomposition given Boltzmann-Shannon entropy, where in general formula to set the integral.

$$\mathrm{H}(X) = \sum_i \mathrm{P}(x_i)\mathrm{I}(x_i) = -\sum_i \mathrm{P}(x_i) \log_b \mathrm{P}(x_i),$$

and

$$\lim_{p \to 0+} p \log(p) = 0.$$

For Information Natural Dynamics [IND] pairwise comparison, the genesis and stigmergetic evolutionary dynamics of the A-group agent **H_A** c.f. for

$$f : \mathbb{R}^n \to \mathbb{R} \textbf{ for } \mathbf{x_i} \in \mathbb{R}^n \textbf{ and } \mathbf{v_i} \in \mathbb{R}^n \tag{13}$$

in Particle Swarm Optimization [PSO].

The variables include function space, gradient, vector and weighting and additionally the stigmergy of the given Macrosystem and subsystem autonomics such as in a Pairwise Breakout Model [PBM].

In associated Fuzzy set logic to determine power externals and singelton mechanics in atomistic Natural Dynamic System [NDS] operations, are often utilized border-pairs or extremals for multi-pair Multilayer Perception-Learning Classification Algorithms [MLP-LCM] [7].

Examples of particle Monadicity in Algorithmic Information Theory [AIT] whence the formula

$$\mathbf{F} : \mathbf{L}(\mathbf{A})^{\mathbf{N}} >> \mathbf{L}(\mathbf{A}) \textbf{where the n-tuple } (R_1, \dots, R_N) \in \mathbf{L}(\mathbf{A})^N \tag{14}$$

indicate the possibilites and types of information physical mechanics for possible variables of the landscape extremals as particles within the min-max parameters [8]. A particle-discrete control function of the node degrees on the evolutionary landscape can therefore be defined where essentially

$$f\ \mathbf{t}(\mathbf{i}) : \mathbf{P}(\mathbf{k})^{\mathbf{n}} >> \mathbf{Kj\ Integral\ o\ t\ k}(\mathbf{t'})\mathbf{Kp\ Sig\ dj}(\mathbf{t})/\mathbf{dt}. \tag{15}$$

To quantify, *Primary extrema* of n-arity or *n-Adicity* [Jonah Lissner] are therefore defined by the Author [Jonah Lissner] for alternatives of pair choices [as monoidal algorithmic circuits in the Complex Adaptive Evolutionary System [CAES]] which can be fractional off the prime polynomial root modulos, from the initial power conditions and therefore generate the discrete information inequalities. These can be demonstrated in Hensellian numbers, and secondly, derivable fractional functions, inherent in any given complex *topos* of an complex adaptive evolutionary system [9].

Nagata defined thusly: A local ring R with maximal ideal m is called **Henselian** if Hensel's lemma holds. This means that if P is a monic polynomial in $R[x]$, then any factorization of its image P in $(R/m)[x]$ into a product of coprime monic polynomials can be lifted to a factorization in $R[x]$ [10].

5.4 Complex adaptive evolutionary system: networks

The network of circuits then form the basis for Complex Network Systems [CNS] from Simple networks $L \propto \log N$ and adaptive complex or dynamic systems and increasingly complex or quantum probability mechanics.

Scale-free or Barbasi-Albert Models are utilized to advance the mechanics and hypotheses for Complex Network Systems [CNS] e.g. for

$$P(k) \sim k^{-3}$$

and

$$p_i = \frac{k_i}{\sum_j k_j},$$

exist for give node degree correlations and links at network computations for a generic clustering law of

$$C(k) = k^{-1}. \tag{16}$$

It can be determined in scale-free network nodes.

$$L \propto \log \log N \;\text{ for }\; P(k) \sim k^{-\gamma}.$$

where

$$s(G) = \sum_{(u,v) \in E} \deg(u) \cdot \deg(v).,$$

and dynamically

$$P(k) \sim k^{-\gamma} \text{ with } \gamma = 1 + \frac{\mu}{a_\infty}.$$

where.

$$p(x_i, x_j) = \frac{\delta x_i x_j}{1 + \delta x_i x_j}.$$

6. Conclusion

These Metaheuristics for Global Optimization Algorithms [GOA] are for purpose of achievement of the theoretical completion between two and more nodes on the network landscape, and ultimately the given requirements for the applied electrical grid. This theory can be utilized to derive, add, multiply, subtract, or divide units designated as necessary to accurately define the parameters for control of the electrical grid, and for control of network extremals.

Some theoretical requirements for Power System applications and machine learning algorithm libraries for solving heuristic challenge for power requirements and control on manifolds have been demonstrated:

1. In the definition for innovation in Global Optimization Algorithms [GOA] for Machine Learning in Power Systems the Path-decision or Algorithm is the activated object [*ontesis*] and Algorithmic network is the kind or type of systemic algorithmic operation of object-getting and technology-building [*telesis*], due to computational physical plasticity conditions and relevant criteria.

2. Furthermore the network theory meaning of Path-decision or Algorithm and the computational landscape itself as a network, can be defined discretely in terms of multiple avenues and nodes for algorithms of Boolean systems, [e.g. *st-connectivity*] whence it has progressed in weight, mass and velocity of the defined Ontology.

3. Path-decision or *Algorithm programmes* in Computational Sciences, by modules of Alphanumeric Symbols/Characters as *Power Systems of Algorithms*, [PSA]-Information Natural Dynamics [IND] or in a macrodynamic method, *Systems*

> *of Utilization* may develop, for the purpose of heuristic advancements based on computational physical references, functions and operations on specific topologies and as treated as given *Computational sequences*.

It is proposed from 3 Rules of Information Physics [3IP] and The 3 General Rules of Macrodynamics [3GRM] for *Unprovable Ideals, Cardinals or Delimitations of Optimization* from origins in *The Rule of Perpetuation of Information Inequalities of Primaries.* These criteria are the basis to utilize previous methodologies of reasoning for contemporary and future new evolutionary algorithmic landscapes in the accretive methods.

This Metatheory develops theoretical agreement for the computational physical basis for a General Global Optimization Field Theory [GGOFT], given the algorithmic requirements of minima and maxima of a set of functions for a given computational surface, to determine roots, stationary and turning points, points of inflection, convexity, and concavity for atomistic qualities of evolutionary landscape extremals and their subsequent geometric values and derivations [11].

Therefore in this dialectic, the *Onts* or Particles in Complex Adaptive Evolutionary System [CAES] and Dynamic Global Workspace Theory-Intelligent Computational System Organization [DGWT-ICSO], can be understood as network gateways in conjunction with nonlinear surfaces, described by *Epistemes,* or Semantical value for given Formulae, Algorithms, Landscape. Their purpose is to build and attempt to game-solve more complex and efficient, workable algorithmic structures for the machine learning algorithm challenges to incremental Global Optimization Algorithm [GOA] regimes [12].

Author details

Jonah Lissner[1,2]

1 Invited Visiting Professor, Center for Mathematical Sciences, Technion — Israel Institute of Technology, Haifa, Israel

2 ATINER Industrial and Electrical Engineering Departments, Athens, Greece

*Address all correspondence to: atiner@atiner.gr

References

[1] "Fermat's Last Theorem", Wikipedia, Website, extracted 2021. https://en.wikipedia.org/wiki/Fermat's_Last_Theorem

[2] The unreasonable effectiveness of mathematics in the natural sciences. Richard Courant lecture in mathematical sciences delivered at New York University, May 11, 1959. Communications on Pure and Applied Mathematics, Volume13, Issue1, February 1960 Pages 1–14.

[3] Meissner, M.; Schmuker, M.; Schneider, G. 2006. "Optimized Particle Swarm Optimization (OPSO) and its application to artificial neural network training". *BMC Bioinformatics*. 7 (1): 125. doi:10.1186/1471-2105-7-125.

[4] "Arrow's Impossibility Theorem", Wikipedia, Website, extracted 2021. https://en.wikipedia.org/wiki/Arrow%27s_impossibility_theorem

[5] Lovbjerg, M.; Krink, T. 2002. "Extending Particle Swarm Optimisers with Self-Organized Criticality" (PDF). *Proceedings of the Fourth Congress on Evolutionary Computation (CEC)*. **2**. pp. 1588–1593.

[6] "Theory of Oscillations", Encyclopedia of Mathematics, Website, extracted 2021. https://encyclopediaofmath.org/wiki/Oscillations,_theory_of

[7] Meissner, M., Schmuker, M. & Schneider, G, 2006. Optimized Particle Swarm Optimization (OPSO) and its application to artificial neural network training. *BMC Bioinformatics* **7**, 125. https://doi.org/10.1186/1471-2105-7-125

[8] Czirok, A., Vicsek, T. 2000. "Collective behavior of interacting self-propelled particles". *Physica A: Statistical Mechanics and its Applications*. Volume 281, Issues 1–4, Pages 17–29. https://doi.org/10.1016/S0378-4371(00)00013-3

[9] Ardema, Rajan, Lang, 1989. "Three-dimensional energy-state extremals in feedback form", Journal of Guidance, Control, Dynamics Volume 12, Number 4

[10] Nagata, Masayoshi, 1953. "On the theory of Henselian rings", *Nagoya Mathematical Journal*, **5**: 45–57, doi: 10.1017/s0027763000015439, ISSN 0027-7630, MR 0051821

[11] R.G. Strongin, Ya.D. Sergeyev, 2014. Global optimization with non-convex constraints: Sequential and parallel algorithms, Kluwer Academic Publishers, Dordrecht. http://www.gbv.de/dms/ilmenau/toc/315801492.PDF

[12] Sergeyev, Ya. D.; Kvasov, D. E.; Mukhametzhanov, M. S., 2018. "On the efficiency of nature-inspired metaheuristics in expensive global optimization with limited budget". *Scientific Reports*. Springer Science and Business Media LLC. **8** (1): 453. doi: 10.1038/s41598-017-18940-4. ISSN 2045–2322. PMC 5765181. PMID 29323223.

Chapter 10

Ultrasonic Detection of Down Syndrome Using Multiscale Quantiser with Convolutional Neural Network

Michael Dinesh Simon and A.R. Kavitha

Abstract

Down Syndrome is a genetic condition that occurs when there is an extra copy of a chromosome 21 in the newly formed fetus. EIF is observed as one of the possible symptoms of DS. But in comparison to the other symptoms like nasal bone hypoplasia, increased thickness in the nuchal fold, EIF is very much less prone to DS. Hence, recommending the pregnant women with EIF to undergo the diagnostic process like amniocentesis, CVS and PUBS is not always a right choice as these diagnostic processes suffer serious drawbacks like miscarriage, uterine infections. This chapter "Ultrasonic Detection of Down Syndrome Using Multiscale Quantiser With Convolutional Neural Network" presents a new ultrasonic method to detect EIF that can cause DS. Ultrasonic Detection of Down Syndrome Using Multiscale Quantiser with Convolutional Neural Network entails two stages namely i) training phase and ii) testing phase. Training phase aims at learning the features of EIF that can cause DS whereas testing phase classifies the EIF into DS positive or DS negative based on the knowledge cluster formed during the training phase. A new algorithm Multiscale Quantiser with the convolutional neural network is used in the training phase. Enhanced Learning Vector Classifier is used in the testing phase to differentiate the normal EIF from EIF causing DS. The performance of the proposed system is analysed in terms of sensitivity, accuracy and specificity.

Keywords: echogenic intracardiac focus, down syndrome, cross-correlation, enhanced learning vector quantiser

1. Introduction

Down syndrome (DS) is the chromosomal abnormality caused in humans when extra genes from chromosome 21 are transferred to a newly produced embryo. It affects the fetal development leading to physical and mental abnormalities. The babies with DS have a distinct appearance than normal babies. Some of the DS victims are shown in **Figure 1**. The babies suffering from DS are likely to have retardation in growth and the mental problems. DS, in general, cannot be cured; the simplest way to avoid the babies with DS is to detect the fetus with DS and prevent it from being born.

DS is not a rare phenomenon and the occurrence of this phenomenon is eventually improving throughout the world. Patterson [1] in his work "Molecular genetic

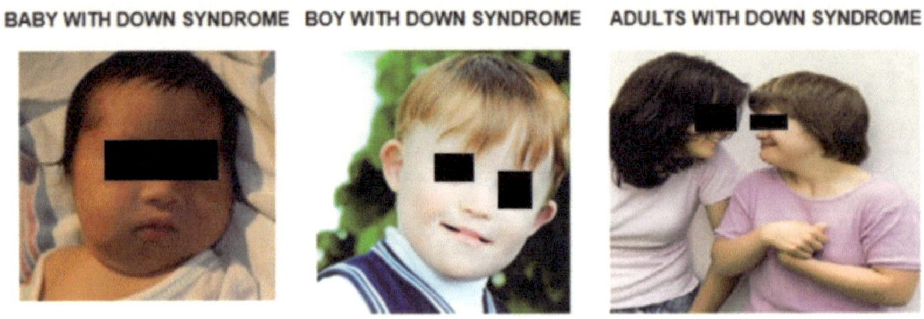

Figure 1.
Victims of DS (courtesy: www.womens-health-advice.com).

analysis of Down Syndrome" states that more than 1 in 1000 neonates has DS. A stats report [2] shows that i) the risk of DS in the global birth rate is 0.00125% ii) women in the age of 30 and below have 0.001% risk of DS iii) women in the age of 45 have 0.022% risk of DS. The abortion rates in DS-affected pregnancies [3] have amplified to 67–92% in the United States and Europe.

In medical world, ultrasonographic markers like nuchal fold, nasal bone hypoplasia and femur length are often seen as a symptom of DS. Nuchal fold is a skin fold noted at the backside of neck in the fetus during the second trimester. Increased thickness in the nuchal fold is observed as the most sensitive symptom of DS. A comparison of the nuchal fold between the normal fetus and DS affected fetus is shown in **Figure 2**.

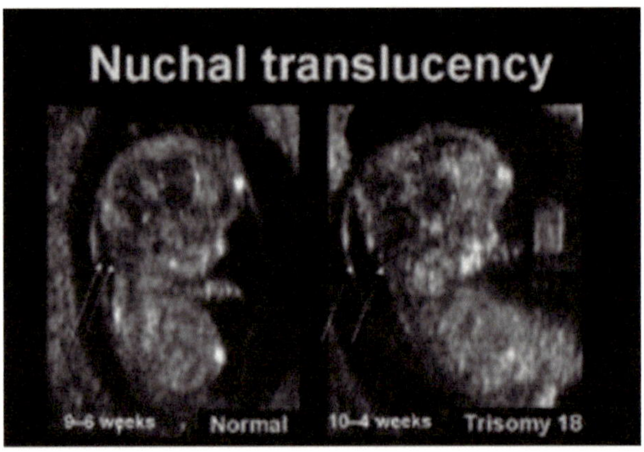

Figure 2.
Comparison of nuchal fold between normal fetus and DS related fetus (courtesy: http://babysmith2014.blogspot.com).

Nasal bone hypoplasia is a condition where the nasal bone of the fetus appears to be very small. It has to be noted that 70% of DS [4] fetuses have no nasal bone or smaller nasal bone. A comparison of nasal bone analysis between the normal fetus and DS related fetus is shown in **Figure 3**.

Femur length is the measure of the longest bone in the fetus. Shortening of femur length is considered as the symptom of DS. Recently, Echogenic Intracardiac Foci (EIF) had been considered as a new symptom of DS. Many clinical researches affirm that the presence of EIF is considered as a potential indication of DS. An EIF is named as an intense white shape that is found in the heart of the fetus. It can be examined through the ultrasound. An article "Ultrasound Findings" expresses that EIF is seen in around 1 out of each 20 or 30 pregnancies (∼3–5%). EIF is associated with 12% of foetuses with DS [5] and biventricular EIF suffers higher risk for

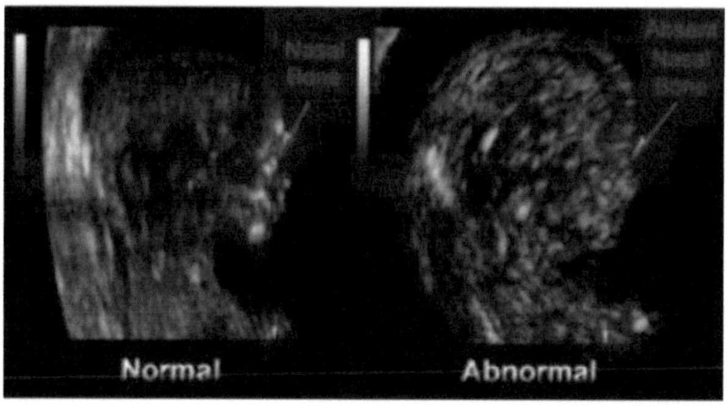

Figure 3.
Comparison of nasal bone between normal fetus and DS related fetus (courtesy: http://babysmith2014.blogspot.com).

aneuploidy. If an EIF is detected in an ultrasound image, the pregnant women with EIF fetus feel dreadful mental pressure fearing that they would deliver DS affected baby [2]. This psychological weight caused in the pregnant ladies can undesirably affect the fetus. To cut the down the psychological weight of the pregnant ladies and to guarantee that the fetus is unaffected with DS, women with EIF symptoms are prescribed to go through chorionic villus sampling and amniocentesis.

Amniocentesis, Percutaneous umbilical blood inspecting and Chorionic villus sampling methods are the most non-ultrasonic method for detecting DS. Amniocentesis is a method of collecting a little amount of the amniotic fluid that surrounds the fetus and analysing it for trisomy 21. It is carried in the light of ultrasonic guidance. It is advised after the 15th week of pregnancy. This method has the high risk of leaking amniotic fluid, miscarriage, needle injury to fetus and infection transmission. Percutaneous umbilical blood sampling (PUBS) is a method of collecting the blood from the umbilical cord and tests it for chromosomal defects. It is performed in the 18th week of the pregnancy. PUBS have a higher risk of miscarriage. Chorionic villus sampling (CVS) is a method of analysing the chromosomes in the cells taken from the placenta. CVS is performed between 9th and 14th weeks of the pregnancy. This method suffers from the drawbacks like miscarriage and uterine infection problems.

As the relativity of EIF with DS is very low, it would not be advisable to request women with EIF fetus to undergo Amniocentesis, CVS and PUBS. There is a great need to coin a new mechanism that can differentiate DS related EIF fetus from the normal EIF fetus through ultrasound testing. This research aims at bringing a new invention that can detect DS based on EIF through an ultrasound scan.

The major contributions of this chapter are:

- A new medical parameter EIF is used for the detection of DS
- A new image segmentation algorithm Multi-scale Quantized Convolution Neural Network (MSQCNN) is developed and used for accurately detecting and extracting the EIF in the ultrasound fetal images
- Cross-Correlation Technique (CCT) is employed to confirm DS by analysing the nasal bone hypoplasia in the training phase
- A new supervised classification scheme Enhanced Learning Vector Quantiser (ELVQ) is utilised to differentiate the normal EIF from the EIF related to DS.

Introduction of the problem statement, the necessity for this research and the challenges present in this research had already been well discussed in section 1. The rest of this chapter is structured as follows. A complete view of related studies is provided in section 2. The proposed methodology is elucidated in section 3. The test results and comparisons are elaborated in section 4. Future research directions and conclusion are briefed in section 5 and section 6.

2. Related works

This section provides a detailed view of the emerging researches performed on the identification of DS utilising ultrasound. Rebecca et.al [6] performed analyses to decide the exactness of second-trimester ultrasound in identifying DS in fetuses. The malformation in the structure and ultrasonographic markers are the baseline factor behind this research examination. The experimental outcomes demonstrated that the ultrasonographic markers without related morphological abnormalities could not differentiate between the normal fetus and the fetus with DS. This research made a noteworthy observation that placing the markers as a deciding factor to suggest amniocentesis can lead to a huge number of fetal losses.

Lauren Lynch et.al [7] brought forth the new ultrasonic parameter nuchal fold that was highly informative for detection of DS than the other parameters like biparietal diameter and femur length. Patrick Rozenberg et.al [8] recommended the inclusion of the first-trimester ultrasound scan for the diagnosis of DS in addition to the second-trimester ultrasound. Detection of the ultrasonographic marker like nuchal fold, biparietal diameter and femur length is a challenging task due to the presence of speckle noise in an ultrasound image. Detection of these markers requires highly skilled sonologists, obstetricians, and fetal medicine professionals [2]. This led to the paradigm shift for the automated diagnosis of DS from the ultrasound images.

Cuckle et al. proposed a genetic scan mechanism called nuchal translucency scan [9] to assess the risk of the fetus suffering from DS based on the nuchal fold. NT scans works based on the principle that "fetus with DS will accumulate more fluid at the end of the neck". NT scan is always performed between 11 to 13 weeks of the pregnancy. As the risk of Aneuploidy and DS increases exponentially with the thickness of NT thickness [10], there has been wide growth in the mechanism of semi-automated methods for detecting DS based on the nuchal fold.

Yinhui Deng et al. [11] have presented a systematic structural model for the automated detection of nuchal translucency region. Anzalone et al. [12] automated the measurement of nuchal translucency from the ultrasound image. Nirmala. S and Palanisamy. V [13] proposed the identification of nuchal translucency utilising the imaging methods like mean shift analysis and canny operators. The thickness of nuchal translucency was computed utilising blob analysis. Sonia. R and Shanthi V [14] performed the morphological operations and Otsu thresholding for the partitioning and calculating the region of nuchal translucency.

Lai K Wee et.al [15] utilised the neural network to figure out the area of nuchal translucency. The boundary region of nuchal translucency layer is detected utilising an instinctive computerised algorithm. After the identification of the boundary region, the optimum thickness of the region is computed based on intensity continuity and edge strength. Yan Du et.al [16] performed fetal karyotype assessment among Chinese ladies and found that nonappearance of nasal bone in the second trimester has the high risk of DS in the fetus. Iliescu Dominic-Gabriel and Drăgușin Roxana-Cristina [17] reviewed various schemes in early hereditary screening and ultrasound assessment and inferred a decision that first-trimester ultrasound assessment can possibly analyse the most of fetal irregularities.

Shafia Shakoor, Humera Ismail and Shama Munim [18] explored the experimental results in fetuses with EIF. The trial was experimented in Pakistan and the outcomes showed that 95.77% of the fetuses were normal, just 4.2% recorded abnormalities in heart and 0% had DS. Aaron B. Caughey, Deirdre J. Lyell et.al [19] assessed the impact of isolated EIF for screening DS. The investigation affirmed the use of EIF as a screening factor may prompt an immense number of amniocenteses and miscarriages to decide a rare DS fetus.

The following observations are evident from the recent researches carried out in this field I) The most often utilised parameters for the identification of DS are nuchal fold, femur length and nasal bone hypoplasia ii) Association of DS and EIF is relatively low iii) There exist a need and necessity for developing a computerised methodology for spotting DS, as the manual diagnosis requires some artifacts.

3. Proposed methodology

Histopathology is the assessment of tissues in the impacted organs of the body. The extensive growth of the computerised image processing has given rise to a new diagnostic methodology termed as computer-assisted diagnosis (CAD). CAD assists the radiologists for disease detection, diagnosis and prognosis prediction. Diagnosis of DS based on EIF is one of the difficult issues that could be tackled by CAD methodology. The efficiency of imaging techniques and CAD can be unitedly employed to substitute the traditional DS diagnostic methods like amniocentesis and chronic sampling.

The proposed system "Ultrasonic Detection of Down Syndrome using Multiscale Quantiser with Convolution Neural Network" uses a medical parameter EIF for the detection of DS. It is designed with the aim of discriminating the normal EIF from DS related EIF. It consists of two phases i) training phase and ii) testing phase. Training phase involves learning and forming the knowledge cluster of EIF related to DS. Testing phase involves classifying the diagnostic ultrasound fetal image as DS positive or negative, based on the knowledge cluster. An architecture diagram of the proposed system is given below in **Figure 4**.

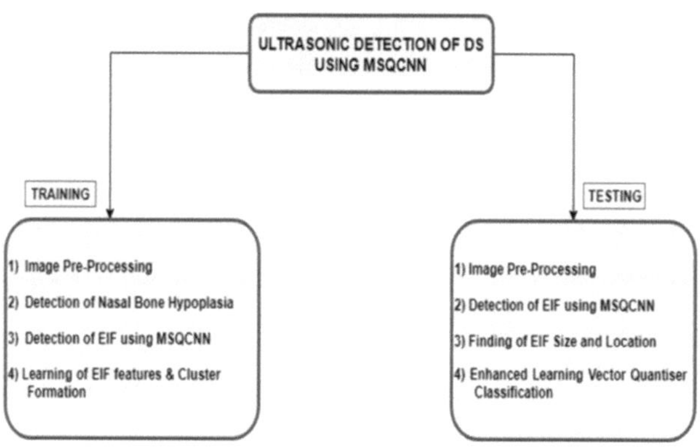

Figure 4.
The architecture of the proposed system.

3.1 Training phase

Not all the fetus that contain EIF result in DS, only few patterns of EIF can cause DS. The aim of this phase is to find the characteristics of DS-related EIF. The training phase includes five steps i) Image Pre-Processing using Neuro-Fuzzy Filter – to eliminate the speckle noise in the ultrasound fetal image that can hinder the detection of EIF ii) Detection of Nasal Bone using CCT - to confirm the presence of DS in the ultrasound fetal image iii) Detection of EIF using MSQCNN – to detect/ segment EIF present in the ultrasound fetal image iv) Learning the features of EIF and cluster formation – to analyse the features of EIF like size, location and form a cluster. The systematic representation of the training phase is shown in **Figure 5**.

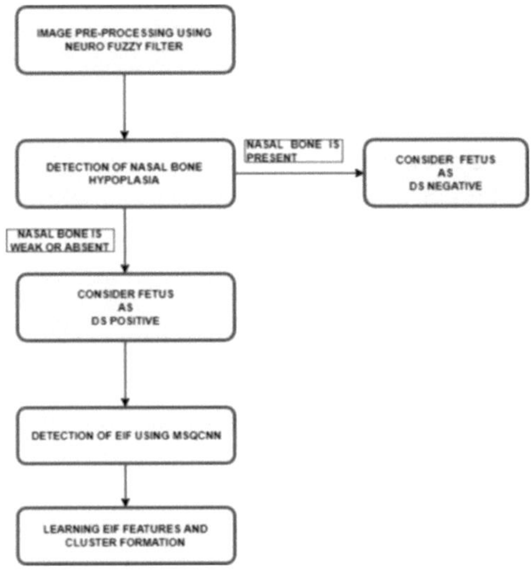

Figure 5.
A schematic representation of the training phase.

3.1.1 Image pre-processing

The ultrasound devices produce high quality, real-time images but very often it is degraded with speckle noise. Speckle noise is a big threat for the quality of the results of image processing. It can produce artificial edges, echoes the patterns in the images that disturb the diagnosis process.

The image pre-processing is the most significant step of the training phase. In pre-processing; a neuro-fuzzy filter that operates based on neuro-fuzzy and evolutionary learning is used to eliminate the speckle noise. The neural network possesses the ability to train. The knowledge for the training is fed into the system in a fuzzy format. Speckle noise is eliminated based on the fuzzy knowledge. A memetic algorithm is utilised to improve neuro-fuzzy filter. The algorithmic description of neuro-fuzzy filter is given below:

Algorithm: Neuro-fuzzy filter
Input: Ultrasound image with speckle noise
Output: Processed image with less noise
Algorithm

- For each ultrasound image X1, there presents 25 membership function (μ)
- Membership function can generate 25 transitional outcomes Y1, Y2….Y25
- Find the weighting factor for each individual membership
- Average the weight of the individual rule's output and form Yo output.

3.1.2 Detection of nasal bone using CCT

Detection of nasal bone is the second step of the training phase. The presence of DS is confirmed by the absence of nasal bone using CCT, as seventy percent of DS foetuses have no nasal bone [4]. CCT is a simple matching algorithm that works based on the correlation. The algorithmic description of CCT is displayed below [20].

Procedure: Cross correlation technique
Input: Pre-processed image
Output: 3D surface plotted graph
Algorithm

1. Start
2. Consider target image S and template image T.
3. Consider the ultrasound image S for which nasal bone has to be detected

$$S'i,j = \frac{1}{mxn} \sum_{i=0,j=0}^{n-1,m-1} S(i+x, j+y) \tag{1}$$

$$\text{where } Si \rightarrow \text{subset of the target image,}$$
$$i \text{ and } j \text{ are coordinates of the template image}$$
$$x \text{ and } y \text{ are coordinates of the target image}$$

4. Consider the ultra sound image T with nasal bone detected

$$T' = \frac{i}{mxn} \sum_{i=0,j=0}^{n-1,m-1} Ti,j \tag{2}$$

$$\text{where } T' \rightarrow \text{subset of template image}$$

5. Derive subset image(I) in T image that include nasal bone area
6. Look for I in S image
7. If I is present in S, obtain the location of I
8. Find cross correlation matrix for the template image and target image

$$\partial = \frac{J(Si+x, j+y-S'i,j)(Tx,y-T')}{\sqrt{J(Si+x-S'i,j)2 * J(Tx,y-T')2}} \tag{3}$$

where $J = \sum_{x=0,y=0}^{n-1,m-1}$ -> summation of co-ordinates in the template image

9. The correlation matrix is converted into surface plotting graph
10. Based on the graph, the maximum value of image correlation will be used for the detection of nasal bone
11. If the peak of the graph is less than 0.35, it can be concluded that nasal bone is absent
12. Stop

3.1.3 Detection of EIF using MSQCNN

Detection of EIF is the third step in the training phase following the image pre-processing and detection of the nasal bone. Training phase continues to the detection of EIF, only when the detection of nasal bone reports that fetal image has no nasal bone or weak nasal bone in it.

MSQNN is used to detect the EIF in the ultrasound images. MSQNN consists of additional components such as acceptor, quantiser with the conventional five layers that include convolution layer 1 and 2, pooling layer 1 and 2, fully connected layer 1 and 2. The output object coming out of FC Layer 2 is led into acceptor. The functionality of acceptor is to compare the object detected in the various scale of the input image and compute the difference between the objects. If the difference between the objects detected at the various scale lies within the threshold, it is treated as the final object. Acceptor computes the difference between the objects using the following formula.

$$\partial = \frac{1}{n} \in \tag{4}$$

where ∂ = Difference between the object detected at scale 1 and scale 2,
Y (i) = Object detected at scale 1,
Y (i + 1) = Object detected at scale 2,
SF = Scale factor at which object is detected.

Acceptor stops forwarding the control to Quantiser if the acceptance criteria are met. Below is the condition for the acceptance criteria to be met.

$$\partial t <= \partial \tag{5}$$

where ∂t = threshold difference between objects detected at various scale (specified by the user).

∂ = actual difference between objects detected at various scale.

If the acceptance criterion is not met, acceptor transfers the control to quantiser. Quantiser performs quantisation by dividing the input image by scaling factor and rounding this value to the nearest integer. The mathematical formulation of quantiser is presented below.

$$X'(u,v) = Round\frac{X(u,v)}{Q(u,v)} \tag{6}$$

The next iteration for object detection starts with the reduced size input image. This iterative process of detecting an object continues until the number of iterations reaches the maximum or the acceptance criterion is met in Acceptor. The final object detected is treated as the EIF present in the ultrasound image of the fetus.

The pictorial representation of Multiscale Quantised Convolution Neural Network is given below in **Figure 6**.

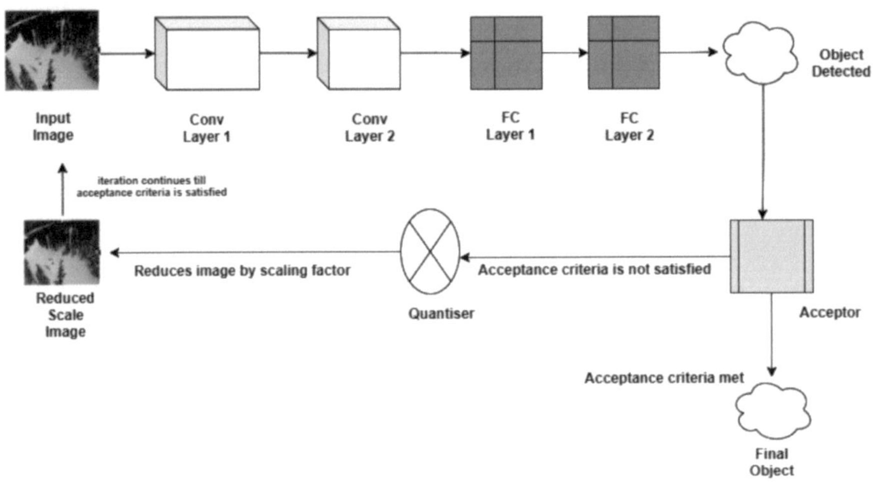

Figure 6.
A pictorial representation of multiscale quantised convolution neural network.

3.1.4 Learning the features of EIF

Learning the features of EIF is the last step of the training phase, it happens just when the nasal bone is absent. The size and spatial coordinates of EIF are stored and clustered based on the closeness of the appearance (2). The similarity parameter utilised for clustering are i) size of EIF and ii) position of EIF specifically left ventricle and right ventricle. These clusters become the knowledge base for ELVQ function of the testing stage. Basically, two clusters are formed in this training phase of ELVQ, one cluster is for DS positive EIF fetus and other is for DS negative EIF fetus.

3.2 Testing phase

Testing phase involves discriminating DS related EIF from the normal EIF. The knowledge attained in the training phase plays the critical role in the determination of DS-related EIF. ELVQ maps the testing image to the cluster map developed in the training phase, by spotting the malignant EIF.

The testing phase contains five steps namely i) Image Pre-Processing using Neuro-Fuzzy Filter – to eliminate the speckle noise in the ultrasound fetal image that can hinder the detection of EIF ii) Detection of EIF using MSQCNN – to detect EIF present in the ultrasound fetal image iii) Find the features of EIF – to find the features of EIF like size and location iv) Enhanced Learning Vector Quantisation Classification – to classify fetus into DS positive or negative based on the knowledge cluster. The flowchart representation of the testing phase is shown in **Figure 7**.

Figure 7.
A flowchart representation of the testing phase.

3.2.1 Enhanced learning vector Quantiser

The first two steps of testing phase are i) Image Pre-Processing and ii) Detection of EIF using MSQCNN are discussed well in the section 3.1.1 and 3.1.3. The third step concentrates on obtaining the features of EIF. Then as the fourth step, ELVQ classification involves.

ELVQ supervised learning classifies the test image into DS positive or negative groups. For a single testing, two steps of ELVQ should be finished. In the initial step, the size of EIF is treated as the weight vector and in the second the location of EIF is treated as the weight vector (2). Enhanced Learning Vector Quantiser differs from the conventional Learning Vector Quantiser in one aspect; LVQ uses Euclidean distance whereas ELVQ uses Manhattan distance function.

Procedure: Enhanced Learning Vector Quantiser
Input: network weights, the learning rate, neighbourhood radius
Output: Classification of zp input vector [21]

1. Set the weights of network, the rate of learning and the neighbourhood radius;
2. while stopping condition(s) not true do
3. for each pattern p do
4. Compute Manhattan distance, dk,p, among input vectors zp and each
   Weight vector uk = (uk1, uk2, · · ·, uKI) as
   dk,p (zp,uk) = |zp − uk| (7)
5. Calculte the output unit ok for which the distance dk,p is the lowest;
6. Update all the neighbourhood weights κk,p using Eq. [7]
   end
7. Update the learning rate;
8. Reduce the neighbourhood radius at specified learning iterations;
9. End

At the end of this testing phase, ELVQ classifies an input EIF fetus image as DS positive or negative.

4. Results and discussions

This section describes the experimental results of the proposed techniques. The proposed system is implemented with software Scilab. It [2] was employed for image and mathematical processing needs. Scilab is open source and freeware that can perform wide numerical computation for engineering and scientific applications.

4.1 Dataset description

As there is no standard dataset for EIF fetus images, the experiment was carried out on the user created the dataset. Nearly 35 2D-ultrasound fetal images were collected from numerous online and offline sources. These fetal images belong to the period of 24–26 weeks. Out of these 35 fetal images, the diagnosis result of DS is known for 25 fetal images (that includes 20 ultrasound fetal images with DS positive and 5 ultrasound fetal images with DS negative).

These 25 ultrasound fetal images are used in the training phase to obtain the characteristics of EIF associated with DS. There were also 10 ultrasound fetal images with EIF with no clue for DS were used in the testing phase. Some of the fetal images with EIF that were used for the experiment were shown in **Figure 8**.

4.2 Experimental results

Training was performed with 25 DS diagnosed images. It was done in five iterations and the learning rate was consistently improving. The training phase results were presented in **Table 1**. The average rate of the learning at the end of training phase was 80%, while the rejection rate was only 20%.

Experimental results of the training phase are given as follows:

Detection of EIF using MSQCNN is shown in **Figure 9**.

DOI: http://dx.doi.org/10.5772/intechopen.96502

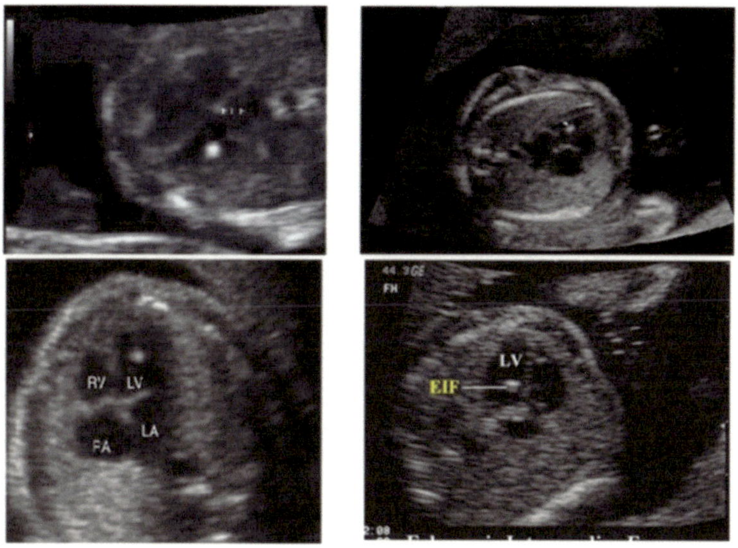

Figure 8.
The dataset of the fetal images with EIF.

Iteration No#	No. Of Input Images	Learning Rate	Rejection Rate
1	10	10	0
2	20	14	6
3	40	34	6
4	80	68	12
5	100	86	14

Table 1.
Training phase results.

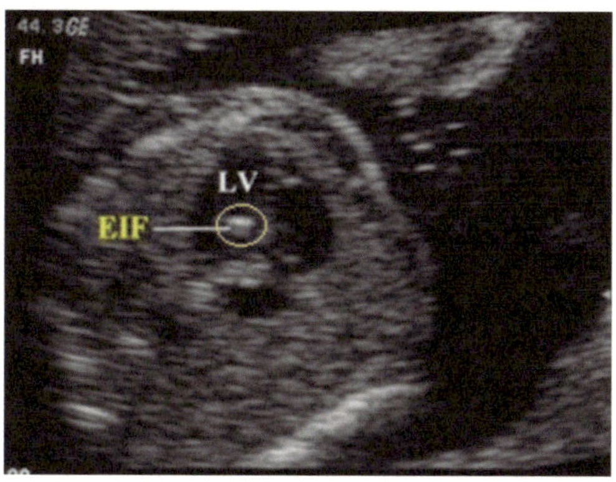

Figure 9.
Detection of EIF.

Detection of nasal bone using CCT in **Figure 10**.

The fetal image (A. Normal) that was rejected and the fetal image (B. Down syndrome) that was learnt in the training phase is shown in **Figure 11**.

The knowledge attained from the training phase about the location and size of EIF is shown in **Table 2**.

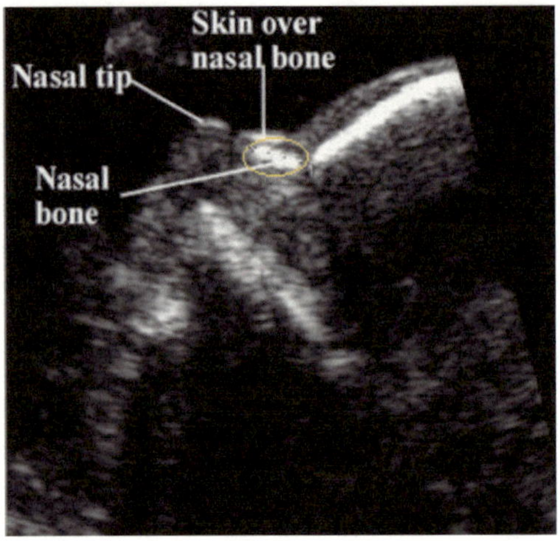

Figure 10.
Detection of nasal bone.

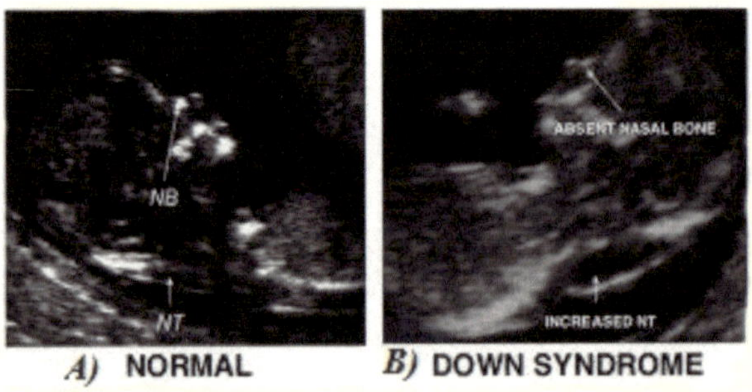

Figure 11.
Results of the training phase.

S.No	Size of EIF	Location of EIF	Other Ultrasound findings	DS Positive
1	> = 5.4 mm	Left Ventricle	Nasal Bone Absence	Yes
2	> = 5 mm	Biventricular	Nasal Bone Absence	Yes

Table 2.
Clustered knowledge from the training phase.

The testing phase was carried out to identify DS in 5 fetal images with EIF is shown in **Table 3**.

Experiment No #	Tested images	Down Syndrome Positive	Down Syndrome Negative
1	50	33	17

Table 3.
Results of the testing phase.

The fetal image with EIF that was analysed for DS in the testing phase and proved to be Down syndrome negative is displayed in **Figure 12**.

The fetal image with EIF that was analysed for DS in the testing phase and confirmed as Down syndrome positive is displayed in **Figure 13**.

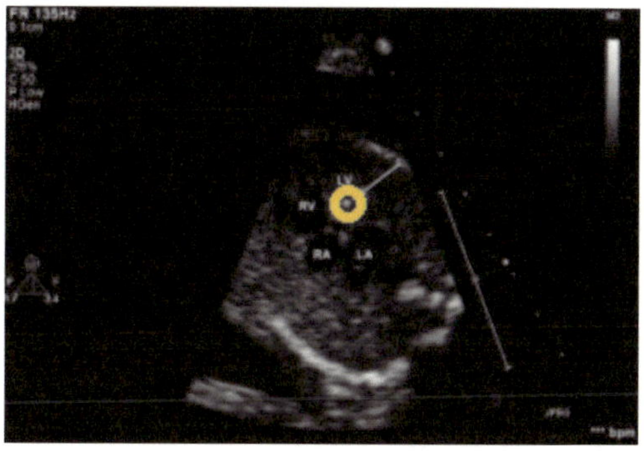

Figure 12.
26 weeks fetus with left ventricular EIF with DS negative in the testing phase.

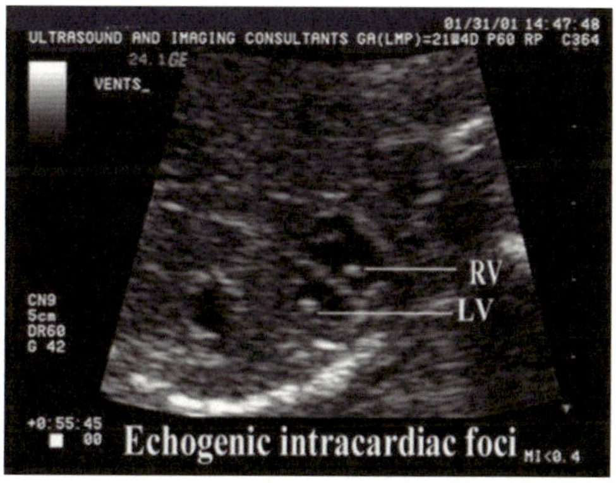

Figure 13.
A fetus with EIF on both the ventricles with DS positive in testing phase. X.

4.3 Performance analysis using evaluation metrics

The performance of DS detection system is evaluated using parameters such as sensitivity, specificity, and accuracy. Sensitivity is the number of true positives that are rightly acknowledged by a diagnostic test [2]. It states how well the diagnostic test is identifying a disease.

$$\text{Sensitivity} = \text{T.P}/(\text{T.P} + \text{F.N}) \tag{7}$$

Specificity is the quantity of the true negatives rightly acknowledged by a diagnostic test [2]. It determines how good the test is identifying normal (negative) condition.

$$\text{Specificity} = \text{T.N}/(\text{T.N} + \text{F.P}) \tag{8}$$

Accuracy is the number of true results, either true positive or true negative, in a population. It measures the degree of veracity of a diagnostic test on a condition.

$$\text{Accuracy} = (\text{T.N} + \text{T.P})/(\text{T.N} + \text{T.P} + \text{F.N} + \text{F.P}) \tag{9}$$

(Note: T.P stands for True Positive, T.N stands for True Negative, F.P stands for False Positive, F.N stands for False Negative)

The performance metrics of the proposed system is compared with the existing system "Automatic Computerized Diagnostic Tool for Down Syndrome Detection in Fetus" [2] is shown in **Figure 14**.

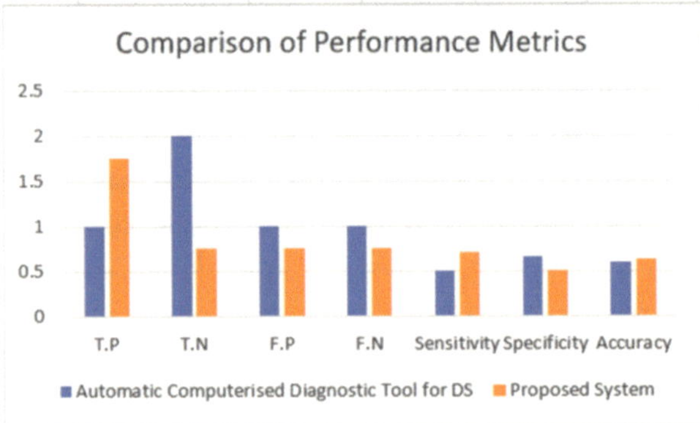

Figure 14.
Comparison with the state of art.

The proposed system showed the better performance than the existing system terms of sensitivity and accuracy, but it showed low specificity than the existing system.

5. Future research directions

The newly designed system was able to clearly differentiate DS related fetus from the normal fetus based on EIF. It was producing very accurate results when operated on the fetal ultrasound images with a single EIF and multiple EIF. In future works, the soft markers like nuchal fold and femur length can be considered as an alternate parameter instead of nasal bone hypoplasia in the training phase.

6. Conclusion

This chapter presented a new idea "Ultrasonic Detection of Down Syndrome using Multiscale Quantiser with Convolutional Neural Network" to detect DS based on EIF in an ultrasonic automated method. The proposed system was intelligent enough to clearly distinguish DS causing EIF from the normal EIF. It attained better results in terms of accuracy, sensitivity, and specificity. In future works, this system can be added as a new feature in the ultrasound fetal scan. It can also serve as an alternate for the conventional DS diagnostics like amniocentesis, PUBS, and CVS.

Acknowledgements

I feel proud and privileged to thank my guide Dr.A.R.Kavitha M.E., Ph.D. for her precious guidance and whole hearted support. I wish to express a sense of gratitude to my lovable English Teacher Mrs. Sucy George M.A., M.A., B.Ed for copy editing of this chapter. Last but not least I wish to dedicate this chapter to my family members (Late Simon – My Father, Mrs. Shanthi Simon - My Mother, Mrs. Gnana

Rita M.A., B.Ed. - My Sibling, Mrs. Pearlin M.E - My Wife) and well-wishers (Mrs. Vimala Gilbert, Rev.Fr.Yesu Antony, Ms.Tamil Ilakika) for their boundless honest support and consistent motivation.

Author details

Michael Dinesh Simon[1]* and A.R. Kavitha[2]

1 Anna University, Chennai, India

2 Department of Computer Science and Engineering, SRM Institute of Science and Technology, Vadapalani Campus, Chennai, India

*Address all correspondence to: ermichaeldinesh@gmail.com

References

[1] Patterson D. Molecular genetic analysis of Down syndrome. Human Genetics. 2009;126(1):195–214. DOI: 10.1007/s00439-009-0696-8

[2] Simon Michael, A.R Kavitha. Automatic Computerized Diagnostic Tool for Down Syndrome Detection in Fetus. In: Nilanjan Dey, Amira S. Ashour, Harihar Kalia, R.T. Goswami, Himansu Das. Histopathological Image Analysis in Medical Decision Making. IGI Global;2019. DOI:10.4018/978-1-5225-6316-7

[3] Mansfield C, Hopfer S, Marteau T M. Termination rates after prenatal diagnosis of Down syndrome, spina bifida, anencephaly, and Turner and Klinefelter syndromes: A systematic literature review. Prenatal Diagnosis. 1999;19(9):808–812. DOI:10.1002/(sici)1097-0223(199909)19:93.3.co;2–2

[4] Cicero S, Rembouskos G, Vandecruys H, Hogg M, Nicolaides K H. Likelihood ratio for trisomy 21 in fetuses with absent nasal bone at the 11–14-week scan. Ultrasound in Obstetrics and Gynecology. 2004;23(3):218–223. DOI: 10.1002/uog.992

[5] Anderson N, Jyoti R. Relationship of isolated fetal intracardiac echogenic focus to trisomy 21 at the mid-trimester sonogram in women younger than 35 years. Ultrasound in Obstetrics and Gynecology. 2003;21(4):354–358. DOI: 10.1002/uog.89

[6] Smith-Bindman Rebecca, Wylie Hosmer, Vickie A. Feldstein, Jonathan J. Deeks, James D. Goldberg. Second-Trimester Ultrasound to Detect Fetuses With Down Syndrome. Jama. 2001;285 (8):1044–1055. DOI:10.1001/jama.285.8.1044.

[7] Lynch Lauren, Gertrud S. Berkowitz, Usha Chitkara, Isabelle A. Wilkins. Ultrasound Detection of Down Syndrome: Is It Really Possible?. Journal of Diagnostic Medical Sonography. 1989;5(4):217. DOI:10.1177/875647938900500445.

[8] Rozenberg Patrick, Laurence Bussières, Sylvie Chevret, Jean Pierre Bernard, Lydia Malagrida, Howard Cuckle, Chantal Chabry, Isabelle Durand-Zaleski, Laurent Bidat, Isabelle Lacroix, Max Moulis, Marc Roger, Marie Christine Jacquemot, Jean Philippe Bault, Philippe Boukobza, Patrick Boccara, Francois Vialard, Yves Giudicelli, Yves Ville. Screening for Down Syndrome Using First-trimester Combined Screening Followed by Second-trimester Ultrasound Examination in an Unselected Population. American Journal of Obstetrics and Gynecology. 2006;195 (5):1379–1387. DOI:10.1016/j.ajog.2006.02.046.

[9] Howard Cuckle, Peter Benn, Dave Wright. Down Syndrome Screening in the First and/or Second Trimester: Model Predicted Performance Using Meta-Analysis Parameters. Seminars in Perinatology. 2005;29(4):252–257. DOI: https://doi.org/10.1053/j.sempe ri.2005.05.004.

[10] Sonek J. First-trimester ultrasonography in screening and detection of fetal anomalies. American Journal of Medical Genetics Part C: Seminars in Medical Genetics. 2007;145 C(1):45–61. DOI:10.1002/ajmg.c.30120

[11] Y. Deng, Y. Wang, P. Chen. Automated detection of fetal nuchal translucency based on hierarchical structural model. In: Proceedings of IEEE 23rd International Symposium on Computer-Based Medical Systems (CBMS); Perth WA; 2010. p. 78–84

[12] Anzalone A, G. Fusco, F. Isgro, E. Orlandi, R. Prevete, G. Sciortino, D. Tegolo, C. Valenti. A System for the

Automatic Measurement of the Nuchal Translucency Thickness from Ultrasound Video Stream of the Foetus. In: Proceedings of the 26th IEEE International Symposium on Computer-Based Medical Systems; 2013.p. 239–44. DOI:10.1109/cbms.2013.6627795.

[13] S.Nirmala, V.Palanisamy. Measurement of Nuchal Translucency Thickness in First Trimester Ultrasound Fetal Images for Detection of Chromosomal Abnormalities. International Journal of Computer Science and Information Security. 2009; 6(3):101–106.

[14] Sonia.R , V. Shanthi. Early Detection of Down Syndrome Marker by Measuring Fetal Nuchal Translucency Thickness from Ultrasound Images during First Trimester. Indian Journal of Science and Technology. 2016;9(21):1–6. DOI: 10.17485/ijst/2016/v9i21/95174.

[15] Wee Lai K, Too Yuen Min, Adeela Arooj, Eko Supriyanto. Nuchal Translucency Marker Detection Based on Artificial Neural Network and Measurement via Bidirectional Iteration Forward Propagation. WSEAS Transactions on Information Science and Application. 2010;7(8):1025–1036.

[16] Du, Yan, Yunyun Ren, Yingliu Yan, Li Cao. Absent Fetal Nasal Bone in the Second Trimester and Risk of Abnormal Karyotype in a Prescreened Population of Chinese Women. Acta Obstetricia Et Gynecologica Scandinavica. 2017;97(2): 180–186. DOIT:10.1111/aogs.13263.

[17] Cristina Drăgușin Roxana. Prenatal Diagnosis of Down Syndrome. In: Iliescu Dominic Gabriel. Advances in Research on Down Syndrome. INTECH Open Access Publisher;2011.p.75–92. DOI: 10.5772/intechopen.71064.

[18] Shafia Shakoor, Humera Ismail , Shama Munim. Intracardiac echogenic focus and fetal outcome – review of cases from a tertiary care centre in Karachi, Pakistan. The Journal of Maternal-Fetal & Neonatal Medicine. 2013;26(1):2–4. DOI: 10.3109/14767058.2012.703724

[19] Caughey, Aaron B, Deirdre J. Lyell, Roy A. Filly, Eugene A. Washington, Mary E. Norton. The Impact of the Use of the Isolated Echogenic Intracardiac Focus as a Screen for Down Syndrome in Women under the Age of 35 Years. American Journal of Obstetrics and Gynecology. 2001;185(5):1021–1027. DOI:10.1067/mob.2001.117674.

[20] Winter Thomas C., Amy M. Anderson, Edith Y. Cheng, Cathy A. Komarniski, Vivienne L. Souter, Stefanie B. Uhrich, David A. Nyberg. Echogenic Intracardiac Focus in 2nd-Trimester Fetuses with Trisomy 21: Usefulness as a US Marker. Radiology. 2000;216(2):450–456. DOI:10.1148/radiology.216.2.r00au32450.

[21] Engelbrecht Andries P. Unsupervised Learning Neural Networks. In: Computational Intelligence: An Introduction. John Wiley & Sons;2007. p. 55–72.

Chapter 11

K-Means Efficient Energy Routing Protocol for Maximizing Vitality of WSNs

Bouakkaz Fatima, Ali Wided, Guemmadi Sabrina and Derdour Makhlouf

Abstract

The progress of wireless communication and microelectronics create wireless sensor network, which is a very important field of research, The utilization of Wireless Sensor Network is growing and have a diversity applications like Military applications, Agriculture, Health care, Medical monitoring. The main issue of WSN is energy consumption, where prolonged network lifetime, is important necessity. From the solution proposed the Clustering with k-means is a successful technique for achieving these goals. This work is adaptation of one of the most famous protocol in WSN witch is Low Energy Adaptive Clustering Hierarchy (LEACH) in the clustering phase where the choice of number of clusters and their CHs.sing the k-means method and the distance between nodes and residual energy. Clustering k-means given a best partition with cluster separation. This chapter regulated as below, in section two we discussed related work used k-means to improved vitality of WSN. In the next section, we introduce the proposed adaptation protocol. The simulation resultsusing MATLAB have shown that the proposed protocol outperforms LEACH protocol and optimizes the nodes energy and thenetwork lifetime.

Keywords: wireless sensor networks, clustering, routing protocol, low energy consumption, K-means, LEACH

1. Introduction

Wireless Sensors Network (WSN) is composed of vast number of sensor nodes for sensing data. Nodes are deployed densely either in (WSN) itself, or somewhere near to it [1]. Although the sensors nodes are low in cost, they have limited energy [2].

Figure 1 below displays Wireless Sensor Network architecture.

"Clustering" is technique very efficient to prolong lifetime of network, by reducing energy consumption. Clusters are easy to manage as compared to large whole network. In this chapter we propose a new routing protocol based of K-Means mechanisms in phase of clustering and used the same transmission principe of Low Energy Adaptive Clustering Hierarchy (LEACH) protocol in each cluster. Cluster Head (CH) is elected the node having the higher residual energy and minimum distances from k-means of each cluster.

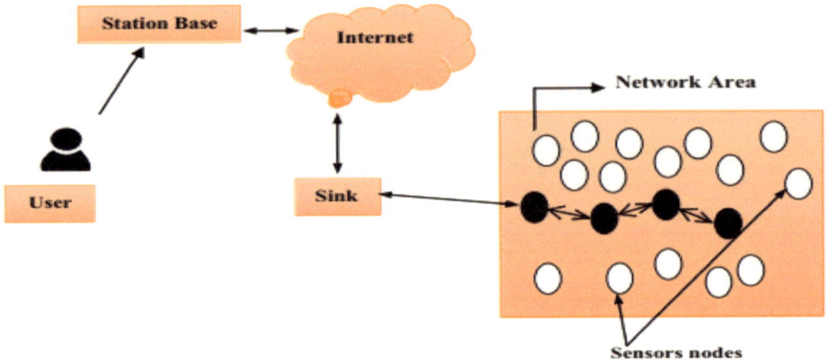

Figure 1.
A wireless sensors network architecture [3].

2. Related work

2.1 The routing protocol low energy adaptive clustering hierarchy (LEACH)

Low Energy Adaptive Clustering Hierarchy LEACH is considered to be the first cluster-based hierarchical routing protocol proposed by Heinzelman et al. As one of the most popular hierarchical routing algorithms for wireless sensor networks [4]. The LEACH protocol assumes equal residual energy from the sensor when starting the network. The life of the network is then divided into towers by a choice of CH. However, each cycle consists of two phases: an initialization phase and a transmission phase (**Figure 2** below displays LEACH cluster formation).

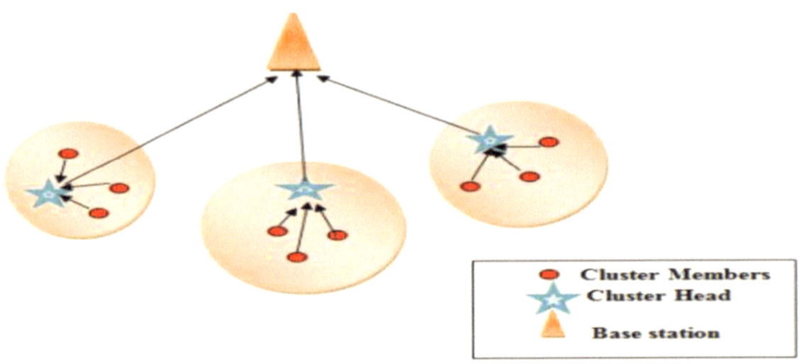

Figure 2.
LEACH cluster formation [5].

2.2 Clustering with K-means

We distinguish several types of clustering techniques: The most used are Partitioning, Hierarchical, Density and Grid based. These algorithms try to decompose the set of nodes into a number of disjoint clusters. The problem is how to select the Cluster Head (CH) and how to manage the clusters. There are many methods of partitioning clustering. The most famous ones are: Fuzzy C-Means, K-means, K-medoids and Partitioning Around Medoids (PAM). The authors [6] surveyed and summarized all the partitioned clustering protocols in WSNs. The next table (**Table 1**) examined a list of routing protocols in wireless sensor networks (WSNs) that uses partitioning clustering with the "K-MEANS" method.

The K-MEANS algorithm is presented as one of the simplest non-supervised learning algorithms that solve clustering problems, developed by J.B.Mac Queen in 1967. In this method we assign a point in each group where the center of this group is closest (centroids). The center is the average of all the points of the group and its coordinates are the arithmetic average of each dimension, together with all the points of the group, which means that each cluster is represented by its center of gravity. **Figure 3** below shows the flowchart of K-means algorithm.

Ref	Characteristics	Intra-cluster	inter–cluster	CH choice	Compared with	Advantages
[7]	k-means used to implement an optimal number of clusters.	One -hop	One -hop	Nœud près du centre de gravité	Clustering traditionnel	• Evolving. • Optimal number of clusters. • Increased net-work life.
[8]	KPSO/KGA	One -hop	One -hop	K-means/ Distance euclidienne	k-means	• Reduces energy con-sumption. • Optimal number of clusters. • Increased net-work life.
[9]	Unique node ID refers to the Euclidian distance between nodes and center of gravity.	One -hop	One -hop	Midpoint algorithm mathematical format	k-means	• More efficient and correct than k-Means for large clusters.
[10]	Combined (K means, Davies Bouldin Index, Gaussian elimination)	multi -hop	One -hop	Euclidian distance residual energy	/	• Better per-formance.
[11]	EBRP	One -hop	One -hop	K-means++ Euclidian Distance	k-means	• Better than k-Means for large clusters.
[12]	Algorithm genetic	One -hop	One -hop	K-means/ Euclidian Distance	GA	• Optimal solution.

Table 1.
The different routing protocol uses k-means in WSNs [3].

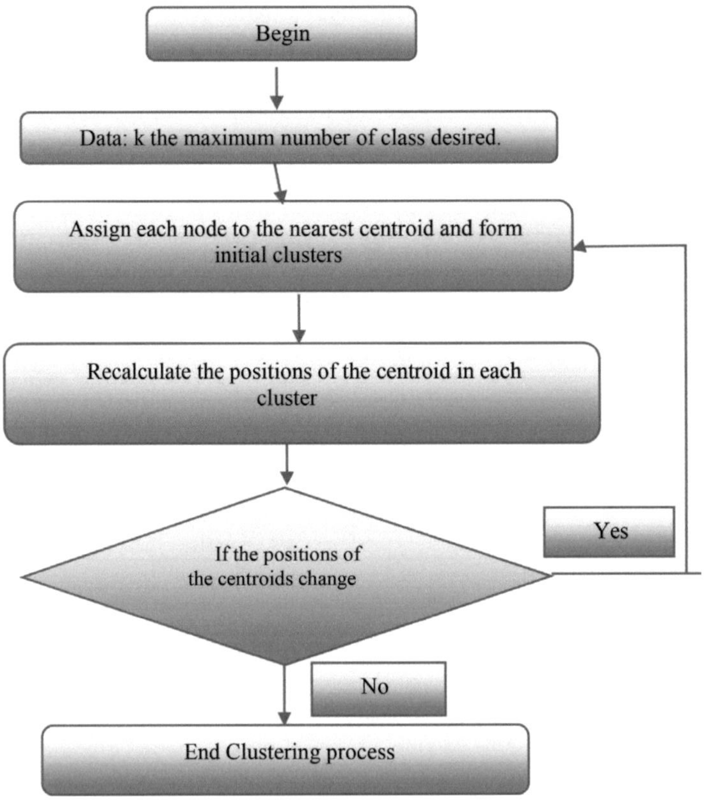

Figure 3.
Flowcharts of K-means algorithm.

3. Proposed system

Our protocol is an adaptation of Low Energy Adaptive Clustering Hierarchy (LEACH) protocol that offers an improvement in the clustering procedure.

In our adaptation the random clustering of LEACH will be changed with the K-means clustering algorithm. This adaptation have been improve the clustering allocation and cluster features and generate energy efficient clustering to increase the life of WSNs.

The use of the K-means algorithm as a clustering technique for cluster formation ensures perfect clustering and reduces overheads when the CHs reelection.

This section presents the configuration of the proposed LEACH protocol adaptation, which consists of two phases: the initialization phase and the transmission phase. **Figure 4** illustrates the two phases of proposed adaptation.

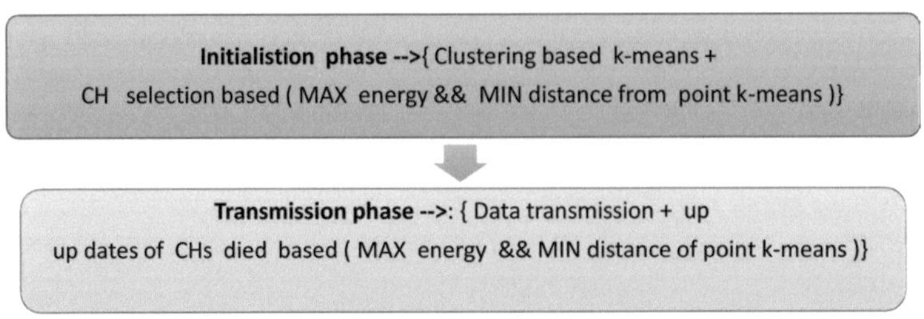

Figure 4.
The two phases of the proposed adaptation.

1. **The initialization phase:** nodes are randomly distributed in the network area, after the clustering process with the K-means method begins, the choice of k-CH is made in this phase using the maximum energy and the minimum distance from k-means distance to choose the CH of each cluster.

2. **The transmission phase:** the nodes of each cluster begins sending collected data to their own cluster head CH, after some iteration, (if the CH energy of cluster <=Min Energy), a CH update procedure will begin among the alive nodes belongs to the cluster using the same parameters of choice new CH(Min distance and Max energy) as the beginning in the initialization phase.

Figure 5 below illustrates the flowchart of operation of the proposed adaptation.

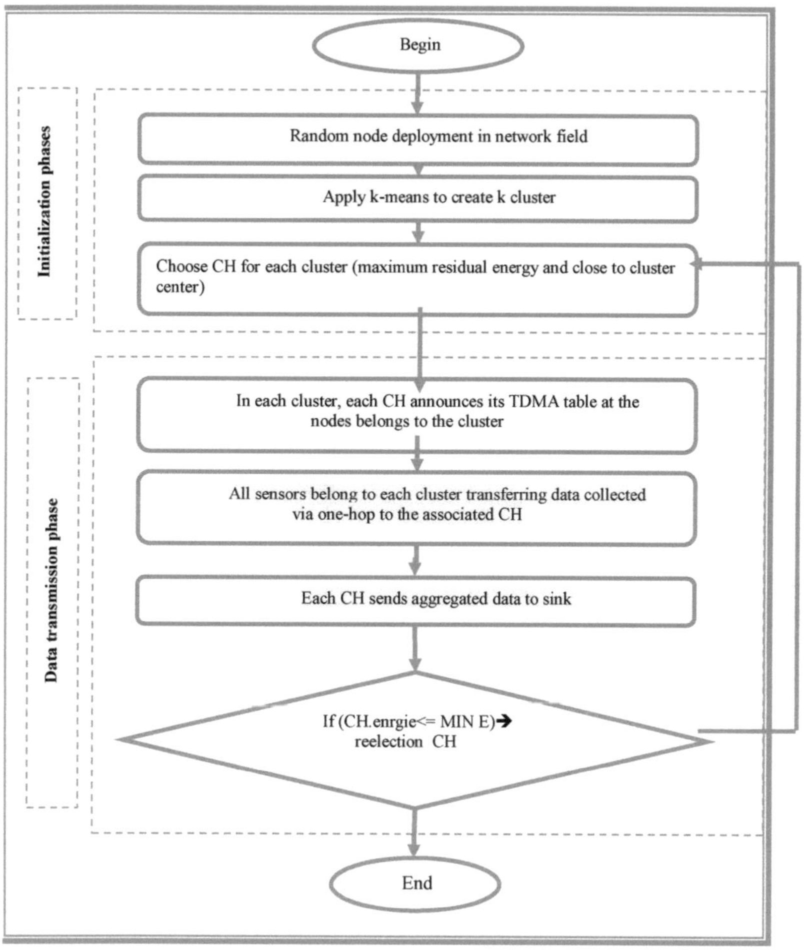

Figure 5.
The proposed adaptation using clustering with k-means method.

4. Experiments and performance evaluation

4.1 Sensor nodes deployment

For example, 100 sensor nodes are randomly deployed over a 100 m^2 area of interest. The SB is positioned at the coordinates (50 m, 200 m). Initially, there is no CH, so the nodes are all normal type. The **Table 2** shows simulation setting.

Parameter	Value
The size of the network.	10 m * 10 m
La localisation de la station de la base.	(50,200)
The number of nodes	100 N
The initial energy of the nodes	0.12 J
Le nombre de cluster à créer avec K-means	5

Table 2.
The simulation settings.

4.2 Simulation results

In this simulation our experimental model is established with 100 nodes randomly spread over a square surface of 100 m^2 and a base station situated in (50,200) coordinate: **Figure 6** illustrates the simulation phase with k = 7.

After the simulation we compare the performance of the two protocols LEACH and LEACH improvement with K-means by certain metrics.

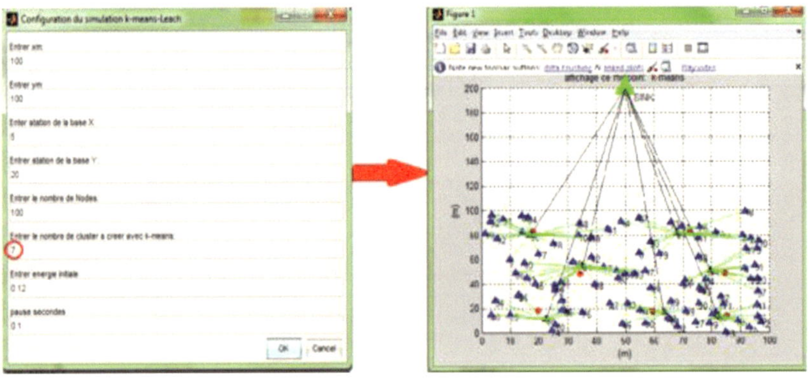

Figure 6.
The phase of clustering with (k = 7).

4.2.1 Residual energy

Figure 7 below represents the residual energy relative to the number of rounds for the two LEACH and K-LEACH protocols. Both protocols showed a gradual decrease in energy and the difference between the two protocols is acceptable. However, K-LEACH show good improvement in lifespan with the same measures.

4.2.2 Lifetime comparison

Figure 8 shows the number of dead nodes in both protocols. In LEACH, there was a very rapid decrease in the number of dead nodes depending on the number of rounds The first dead node in round Number: 97 and the tenth dead node in round 120 and all nodes died Round 441. This is what causes a very short lifespan in LEACH. On the other hand in K-LEACH has a longer lifespan, with a slow decrease in the number of dead nodes according to the round or The first dead node in round Number: 235 and the tenth dead node in round 337 and all nodes died Round 689.

4.2.3 Comparison of CH numbers

Figure 9 shows the number of CHs in both protocols. We find that in LEACH has a high number of CHs which implies high energy consumption on the other hand

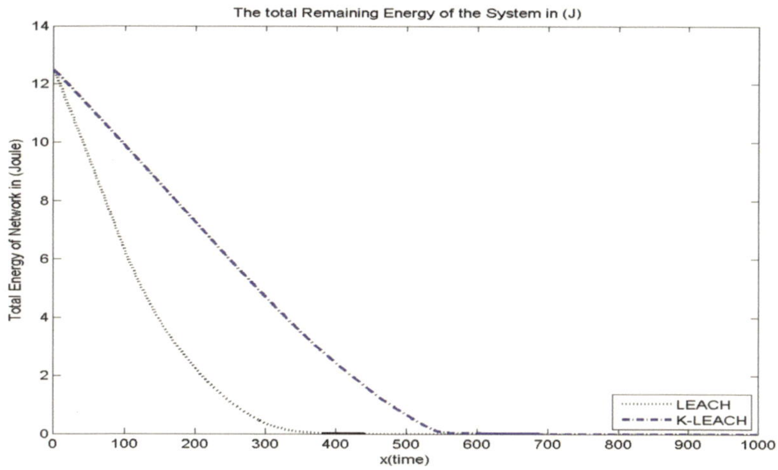

Figure 7.
Residual energy versus rounds (LEACH and K-LEACH).

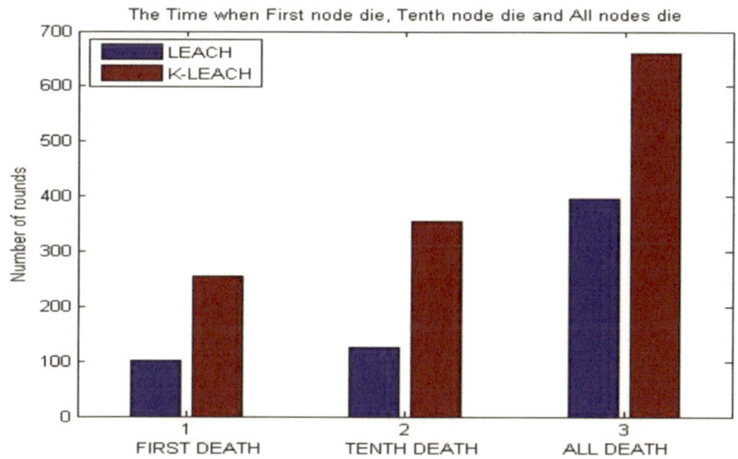

Figure 8.
Number of dead nodes "first, tenth, last" versus rounds.

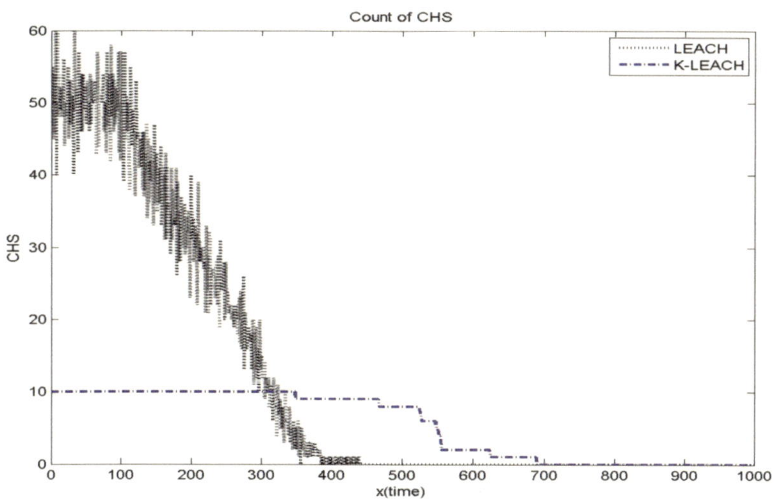

Figure 9.
Number of CHs nodes versus round.

in K-LEACH a fixed number of CH k-10 will die slowly. K-LEACH ensures a good distribution of CH on the network. LEACH presents a huge variation in the number of CH per round this leads to poor network coverage, and affects the overall lifespan of the network.

After comparing the performance of the two LEACH and K-LEACH protocols, we noticed that the our adaptation K-LEACH has several advantages such as: The decrease in energy consumption; A longer lifespan during the simulation with a good distribution of CH thanks to the k-means method.

5. Conclusions

In recent years, Wireless Sensor Networks have been among the most active research themes due to their applications. The main factor limiting a sensor node is energy consumption. The battery-powered sensor nodes must be able to operate at very low Energy consumption.

The cluster-based routing protocols group the sensor nodes to efficiently relay to transmit data to the Sink. Cluster Head CH aggregates the data and sends it to the Sink in the name of the nodes in its group.

The most interesting research problem is how to form the clusters and choose the CHs so that energy consumption and contemporary communication parameters such as latency are optimized. Factors that influence on cluster formation and selection of CHs remain open questions for research.

Simulation results show that adapting the LEACH protocol with the K-Means clustering method that can extend network life and improve energy efficiency, increasing node survival and performance exceeds LEACH's in terms of the amount of data transmitted to the base station and network life.

Author details

Bouakkaz Fatima[1*], Ali Wided[2], Guemmadi Sabrina[2] and Derdour Makhlouf[2]

1 Superior National School of Computer Science (ESI), Algiers, Algeria

2 Tebessa University, Tebessa, Algeria

*Address all correspondence to: f_bouakkez@esi.dz

References

[1] Fatima bouaakaz, makhlouf « State-of-art of grid protocols clustering for wireless sensor networks » ICIST '18 Proceedings of the 8th International Conference on Information Systems and Technologies; March 2018 https://doi.org/10.1145/3200842.3200855

[2] Fatima B, Makhlouf D, Wided A. «Improved Vitality Of Wireless Sensor Network Using Grid Clustering With Multi-Hop Transmission Protocol Routing». Telecommunications and Radio Engineering 79(6):521-532 May 2020. DOI:10.1615/TelecomRadEng.v79.i6.60

[3] Bouakkaz, F., Derdour, M. « Maximizing WSN Life Using Power Efficient Grid-Chain Routing Protocol (PEGCP)». Wireless Pers Commun (2020). https://doi.org/10.1007/s11277-020-07908-9

[4] Y. Yaser, "Routage for Energy Management in Wireless Sensor Networks", PhD thesis, University of Haute Alsace, 08 July 2010.

[5] Kiranpreet kaur, Ridhi Kapoor « Investigation of LEACH Protocol and its Successors in WSN» I. J. Computer Network and Information Security, 2017, 6, 44-52. DOI: 10.5815/ijcnis.2017.06.05

[6] F. Bouakkaz, M.Derdour, W Ali; "Taxonomy of Partitioning Clustering Algorithms in WSNs", Conference on Innovative Trends in Computer Science November 20-21 (2019).

[7] S. Singh and G. Singh. "An Energy Threshold based WSN Clustering Schema using PAM Algorithm". 434 International Journal of Current Engineering and Technology, Vol.5, No.1 (Feb 2015)

[8] A. Sheta and B. Solaiman, "Evolving a Hybrid K-Means Clustering Algorithm for Wireless Sensor Network Using PSO and GAs," Int. J. Comput. Sci. Issues, vol. 12, no. 1, pp. 23-32, 2015.

[9] A. Ray and D. De, "Energy efficient clustering protocol based on K-means (EECPK-means)-midpoint algorithm for enhanced network lifetime in wireless sensor network," IET Wirel. Sens. Syst., vol. 6, no. 6, pp. 181-191, 2016.

[10] R. ELkamel and A. Cherif, "Energy-efficient routing protocol to improve energy consumption in wireless sensors networks," Int. J.Commun. Syst., vol. 30, no. 17, p. e3360, 2017.

[11] L. Li and D. Li, "An Energy-Balanced Routing Protocol for a Wireless Sensor Network," J. Sensors, vol. 2018, pp. 1-12, 2018.

[12] B. Barekatain, S. Dehghani, and M. Pourzaferani, "An EnergyAware Routing Protocol for Wireless Sensor Networks Based on New Combination of Genetic Algorithm & k-means," Procedia Comput. Sci., vol. 72, pp. 552-560, 2015.

Chapter 12

Information Technology Value Engineering (ITVE)

Lukman Abdurrahman

Abstract

Information Technology (IT) has been the main infrastructures in conducting businesses nowadays. IT had firstly emerged to automate business operations, in turn, IT has been able to make business operations more effective due to the nature of IT as a business enabler. Consequently, resource-based business efficiency is also the cornerstone of utilizing IT. Most recently, IT has become OTT (Over the Top), where IT is only the basic layer for conducting online business. In essence, the aforementioned estimations all still rely on qualitative estimates and are not yet based on quantitative estimates. This chapter wants to try to offer an IT value model whose value estimation can be done quantitatively using the Partial Adjustment Valuation (PAV) approach.

Keywords: information technology, business enabler, value, engineering, model, optimization, partial adjustment valuation

1. Introduction

Information Technology (IT) has already transformed into a business enabler and an intentional reason [1] in firms to date. However, the IT presence should get improved administration to cause more values [2] such as effective and efficient business processes, and profit growth [3]. Consequently, IT should replace from a business enabler to a business transformer as per IT ecosystem to convey the approach of IT services [4], in which IT does not only behave a driving instrument but also lets businesses innovate and disrupt customs to revitalize its presence inside the firm. In sequence, this revival will allow the firm to sustain its competitive advantage remains efficient [5].

Additionally, this chapter intends to verify the IT presence in terms of a business transformer in the firm operation [6–8] improving its performance. To do so, it needs a method to engineer the IT position in transforming the business. Also, the method is to show up the IT capitals bringing up more values. In other words, the method should involve a systems engineering viewpoint, which discloses prime thoughts of the systems approach such as holism, synthesis, interrelationships, along with the engineering-project-based estimates of system life cycle and requirements [9]. Likewise, the systems engineering utilizes an engineering design containing problem-solving, alternative solutions, solution selection, detailed model, model guard, and validated model [10].

Also, validating the chapter, the studies on the strength of resources on performance [11–13] turns into an essential theory to analyze IT systems as components of business completeness [14] because the studies emphasize on the resources an organization owns to promote its performance. This is in line with the intent of this chapter that also emphasizes that the business performance runs over IT resources owned [15–17]. Likewise, this chapter applies the Partial Adjustment Valuation (PAV) theory approach in congregating valuation methods among system components as promoted by [18–20]. Equally for the use of PAV in this chapter is the ability to relate between IT resources and the organizational performance mathematically. Thus, it is easier to trace the relationship of each component or subsystem for further synthesis.

Additionally, this chapter has continued the earlier studies addressing the IT value model from the ontological approach towards IT value engineering [21], the IT value model using a variance-based structural equation modeling (SEM) towards IT value engineering [22], engineering IT value in IT-based industries using PAV and RBV (Resource-Based View) approach [23], valuation methodology of IT value in the IT-based business [24], IT value engineering model and its optimum performance [25, 26], and hybrid configuration in IT value models [27]. The investigation leads IT to be valuable resources of the firm to revitalize the IT's role through the IT value engineering model. Formerly, those studies had associated with a number of studies discussing the relationship between IT resources to business performance, such as [11–13, 17–20, 28, 29]. Meanwhile, the IT value defined in this chapter is the added value in the form of a currency, which can also be expressed as the index ratio, generated due to the IT spending presence.

2. Information technology value model study

This chapter problem relates to past studies, which most of them had talked over the relationship between IT and the organization performance. Researchers identified the types of conclusion [8] about the relationship, where numerous conclusions show that the relationship might be positive [12, 13, 30], negative [3], and even neutral [11, 31], see **Figure 1**. Additionally, the positive relationship means that IT has a positive correlation to the organization to increase its business value and at that moment the negative is otherwise. The negative relationship shows in early empirical studies explaining the association between IT investment and organizational performance; they set off the controversy of the IT productivity paradox as Brynjolfsson's (1993) conclusion [32].

In addition, the neutral relationship explicates that IT and business performance have no relationship in between as Strassman (1997) argued that there was no clear relationship between IT investment and a few measures of firm profitability, including return on asset, return on equity, and economic value added [11]. This chapter, further, addresses the positive relationship, although there are unfortunate situations where IT may have a negative impact as well [33].

The earlier studies, especially about the positive conclusion, have not yet talked about how to engineer the value of IT to deliver more benefits to the organization. On one hand, this issue is a complement study to the previous one because the chapter topic is a continual study of the past studies. On the other hand, also, this chapter may strengthen to close the past study, especially in terms of positive relationships between IT and the business performance, therefore, this chapter is to enrich and develop this domain further.

Additionally, the past studies seem like passive research, meaning that the work performed in the IT-equipped organization before. Consequently, the study has been simply conducted in conventional organizations and it has been less hard if it has maneuvered in a planned system. Therefore, the IT value engineering model chapter tries to offer a new approach to studying the role of IT within an

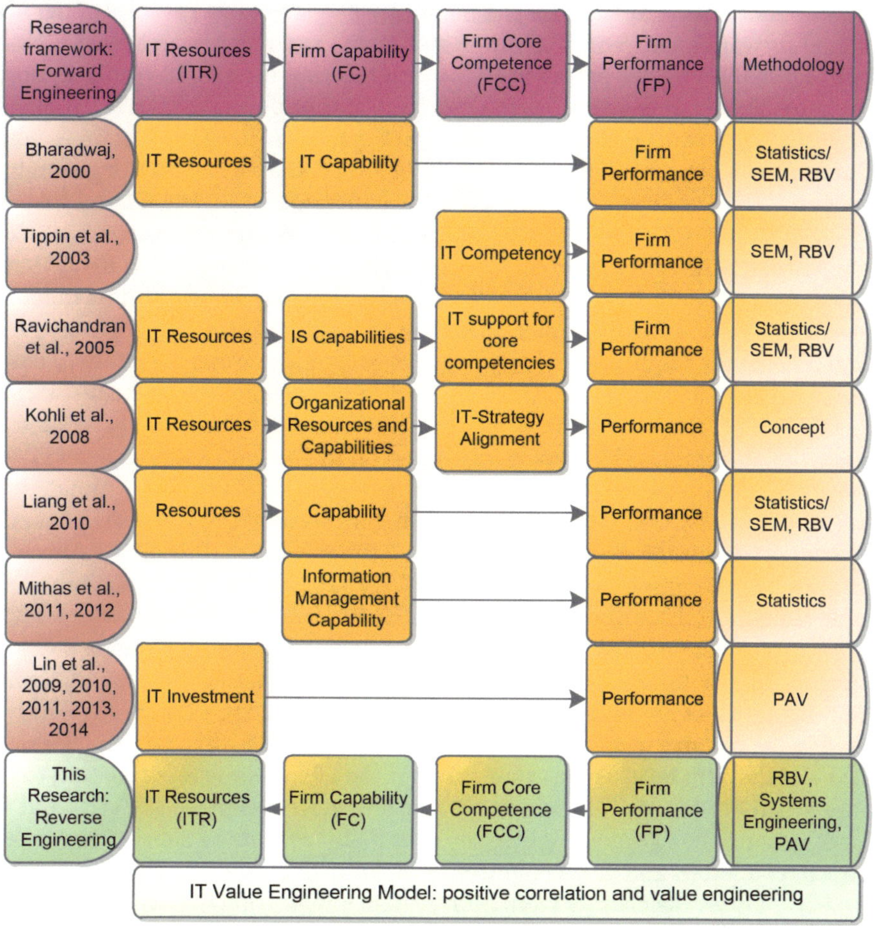

Figure 1.
The state of the art of ITVE.

organization. The approach initiates from RBV theory mentioning that the firm performance should root on the resources the firm owns, as the most famous fundamental theory in studying IT and organizational performance [13]. Then it explores sources of values of IT such that it can carry out a relationship formulation between IT value and the organization.

Figure 1 confirms to place this chapter among the others, which grounded on the earlier ones, nevertheless, with a different approach. For instance, from the subsystems point of views, this chapter solely resembles the Ravichandran's model, which also took in four subsystems, namely IT resources, IS capabilities, IT support for core competencies, and firm performance [12]. Likewise, [34] exhausted four subsystems, although their nomenclature is different from Ravichandran's. As for the other studies such as [11, 13] used three subsystems, namely IT resources, IT capability, and firm performance. However, [35] directly studied to link up between IT competency to firm performance, also [30, 36] simply studied between the data management capability to the performance as well as [18–20, 37, 38] researched the relationship between IT investment and public presentation. This chapter applies four subsystems based on RBV theory with their nomenclature as follows firm performance (FP), firm competence (FCC), firm capability (FC), and IT resource (ITR).

Furthermore, from an engineering point of views, the earlier studies generally exploited forward engineering, which begins from resources towards firm performance, while this chapter proposes reverse engineering for a serial configuration (see **Figure 1**), which begins from the required firm performance, afterwards, estimate the firm core competencies and firm capabilities to get IT resources composition. Also, from a methodological point of views, the earlier studies generally benefitted statistical approach or structural equation modeling (SEM) and RBV approach, excluding [18–20, 37, 38] who used PAV. This chapter appears with a different approach, whereas combining the earlier approaches such as RBV, PAV, and systems engineering at once. In other words, this chapter also carries the different final goal from the earlier ones, in which this chapter is to engineer IT resource about the required firm performance to let it performs at the lower cost.

Additionally, the chapter on the IT value was also conducted by the researchers such as [33] directing that the IT is an integrally part of a system of interrelated organizational factors, [39], who concentrated on the potential and realized IT values estimated by DEA (data envelopment analysis). Similarly, [6] estimated the IT business value by Cobb–Douglass function. Furthermore, [28, 40, 41] generally addressed the IT business value.

Meanwhile, to complete the chapter, the PAV [20] applies to correlate the subsystem input to its output. Additionally, the PAV usually operates with a static speed of adjustment or with a dynamic speed of adjustment in a researched object, on the contrary, in this chapter both the static and the dynamic speed of adjustment work together at a time in the PAV experiment. Likewise, thus far the PAV has applied in the country level study such as [18–20], however, this chapter tries to use the PAV at the firm level as the other study of the IT investment correlation to the firm productivity [8]. As for the chapter analysis, system engineering is to find this chapter because the IT value engineering essentially consists of subsystems such as electrical, computer, and value engineering. Likewise, the system engineering approach analyzes the chapter from both system engineering life cycle and model point of views.

In turn, the chapter result may turn away to become a framework to design an IT-based governing body by looking at several factors either internal or outside factors, including business environment ones. In other words, this chapter is an active chapter using its result, it can plan an organization as well as develop an established organization. Consequently, this chapter encompasses a broader domain of IT-based organization: established and planned systems.

3. IT value engineering paradigm

3.1 Value concept

Discussing a value means that it is addressing usefulness, worth, benefit. Furthermore, the value may disclose if there is an interaction between two or added systems or subsystems, in which one system works with the other one and vice versa, or the system works due to the other systems. Why would the systems mutually function? There is an energy that urges them to work, which is latterly called the value, usefulness, worth, benefit, competitive advantage, or other terms. In other words, this construction can facilitate accomplishing the stage of value creation by benefiting system processes.

There are various types of values such as normative value, realist value, and perceived value. Consecutively, the normative value relates to the required value as planned previously, the realist value pertains to the resulted value that comes from an accomplishment, and the perceived value is what consumer relatively perceives [16]. Additionally, if comprehended from cost management perspectives, the other types of values are the use value, meaning the value of the required function associated with the cost. Afterwards, the cost value, namely all cost values, dedicated to result in the item; the esteem value, means that the value of surplus cost to pay the additional items; the exchange value, namely the value of an item to exchange something else [42, 43].

Moreover, as a fundamental nature of the value definition of this research, the equation definition below bases further studies. This equation technically articulates a value (V) as an index resulted from a function (F) division by cost (C) as proposed by [44] as follows:

$$V = \frac{F}{C} \tag{1}$$

According to the formula, several efforts to bring the value gaining are:

- For a similar function (F), diminish the cost (C) or
- The cost (C) is stable, improve the function (F) or
- The function (F) slightly reduced, the cost (C) significantly decreased or
- The cost (C) a slight increase, the function (F) has increased significantly or
- The function (F) increases while the cost (C) decreases.

It appears that by adjusting function and or cost, the system can control the value to ascend or descend consistent with what the purpose is, although, in practice, there are several considerations to essentially prepare in implementations [44].

3.2 IT value creation

As mentioned, the IT value may come from an estimate of the real worth, utility, or the IT system's significance. This definition does not limit from what the worth, the utility, or the significance come from, thus, it does not prevent the multiple perspectives possibility. There is the stakeholder expectation such this, in turn, it influences the IT value achievement. In substance, value stems from the IT system to support the stakeholder aims attainment. For example, a debit card system that removes the requisite for cashiers to manually count cash may present cashiers with value since it lessens tension on their hands [45].

In addition, to explore the value of IT needs to investigate some scales reflecting these values. Accordingly, the metrics development is a necessity to measure IT values, however, there are certain criteria for the metric development as proposed by [45]. These criteria depart from selected questions that might be considered as follows:

- What is the evidence to evaluate?
- Where must valuation occur?
- When must valuation occur?
- How must valuations be interpreted?

It has been completely recognized that the IT value systems can manifest as a complex system consisting of various subsystems, components, subcomponents, and parts. Furthermore, as measuring the IT value, it is valuable to think about measurements that concentrate on. Definitely, the building of the metrics as a means to evaluate values results in a variety of problems, which necessitates doing so with care [45].

On the other point of view, the IT value study has to involve two sections: (i) IT variable, IT management variable or manifestation, and (ii) endogenous variable with IT economic impact [46]. Doing IT valuation involves complex issues, including social accomplishment so it requires over a period of time. Thus, this study should perform in an inclusive fashion such that the IT value research corresponds to an imperative flow of work that leads to business value. Likewise, there are economic associates of IT and its manifestations, and by itself, the scope of the research should restrict to examine the IT value to engineering it at the organizational level.

In the meantime, the IT resources that are subsequently delivering their capabilities can not create value for themselves within the organization. They need interaction with a business environment such that each will complement one another. As a result, IT infrastructures and organizational factors appear to work in a synergistic way, where these factors are part of the IT-based system consisting of IT human resources and IT management skills, rules, and policies. This is as the organizational system that comprises non-IT human resources and management skills, business procedures, information benefits, affiliation benefits, way of life, organization, and rules. In reality, IT is production machines, therefore, it generates value in the output configuration resulting in benefits due to business processing. In other forms, the value is apparent by itself in the form of process improvements such as saving time, process effectiveness, profitability, such as a higher return on assets, on investment, and consumer surplus such as higher customer satisfaction.

Furthermore, [34] stated that there are numerous factors in terms of the IT value creation chain that is essential and required conditions. Included in these conditions are the IS-strategy configuration, organizational restructuring, business process accomplishment, knowledge sharing, and IT management among others. Accordingly, those are critical in terms of the encouraging of the transformation process and renovation of the effectiveness of IT advantages. Additionally, there are four foremost subjects to demonstrate how IT value is shifting to describe, quantify and show it. The four subjects are (1) value IT-based co-creation, (2) IT embeddedness, (3) information approach, and (4) value extension.

The following stage of the IT value creation should concentrate on the co-creation of value by means of IT instead of IT value itself, further, it is called IT-based co-creation of value. While, the co-creation stands for the thought that (a) IT value cannot manifest in an isolated environment, it is progressively more being formed and accomplished due to actions of numerous parties, (b) value comes from strong joint associations among organizations, and (c) configurations and encouragements for the parties to contribute in and equitably assign emergent values are essential to keeping up co-creation. Moreover, IT embeddedness relates to the condition in which the IT is a central part of the process such that it turns into identical to the product. For example, the IT in a bank's industry of instant credit check is intensely embedded in the loan endorsement process and hard to separate out. In other words, IT embeddedness is a fundamental model that attached to value co-creation, information mindset, and value expansion. Thus, it is plausible that preferred business capabilities drive IT embeddedness. Therefore, the effective convergence of preferred business capabilities and IT capabilities is a precondition "to realizing capabilities among organizations (co-creation), creating information value (information mindset), and ultimately realizing a wide repertoire of value (value expansion)" [34].

3.3 IT value engineering concept

Tohidi (2011) stated that the idea of value engineering is to employ the projects, strengthen accomplishment and diminish costs in all life cycles of the projects. In this case, the lifetime value of the engineering project with the productivity increment can result in the value of the project, namely the output to the input ratio. Hence, the value engineering application is boiling down to performance improvement of inputs and outputs, by applying a theoretical approach of value engineering processes together with project management, project analysis, value analysis, and value management. Additionally, he mentioned that value engineering is constantly dealing with the growth of technology, reducing the unnecessary costs that do not relate to improving the products or services quality. Reducing costs in conventional point of views do not associate with creativity, it only refers to familiarities, feelings, and practices. On the contrary, in value engineering, the usage of knowledge, the problem recognition, the method of problem-solving, the development of the creative solutions could combine to develop comprehensive approaches [47].

Additionally, value engineering is a structured method to investigate the systems function and its completeness in dealing with a systems fundamental function accomplishment at the lowest cost. However, the functions of systems consistently keep up a better performance, trustworthiness, quality, and security [48]. Consequently, the value engineering process arrives at success if it is to discover opportunities to diminish needless costs and at the same time it is able to keep up and raise quality, consistency, accomplishment, and other customer needs on products or services. In terms of IT value, the IT generally boils down on the effectiveness and efficiency of processes, including achieving the best organizational performance. In other words, IT should disseminate value-added advantages through strategic alignment with the organizations.

Essentially, the IT presence within an organization is the norm for the era that is so, there is no one business organization that does not exploit the IT, where the simple difference is the amount of IT capacity. Empirically, such circumstances are something that is unquestionable, but the problem now is how to place the IT position within the business organization with the aim of its presence increasingly contribute enormous weights to the organization performance. Thus far, the IT inclusion in organizations is due to the demands of the times as technology-driven instead of market-driven, it is more emphasis on administrative rather than business development activities. Therefore, this chapter attempts to reposition IT as a means to improve competitive advantages of firms as indicated in **Figure 2**, which appears that IT should set it on the layer where is as the engineering processes culmination that preceded by electrical, computer, software, systems, and a complex system engineering to lead generating an IT value engineering.

Intrinsically, IT value engineering positions the IT at the more well-organized since it can go through an engineering process to create additional significant values as a continuation value generated on the preceding layers (see **Figure 2**). In other words, IT value engineering is the added value due to the engineering of the systems consisting of value, software, computer, electrical engineering. Meanwhile, the organizational environment is a circumstance where a firm should perform its business here, which are competitive forces including risks due to the business activities. The organizational environment should controllable to continue firm's existence in business turbulence to sustain its competitive advantages, which are consisting of six categories, namely cost, differentiation, focus, execution,

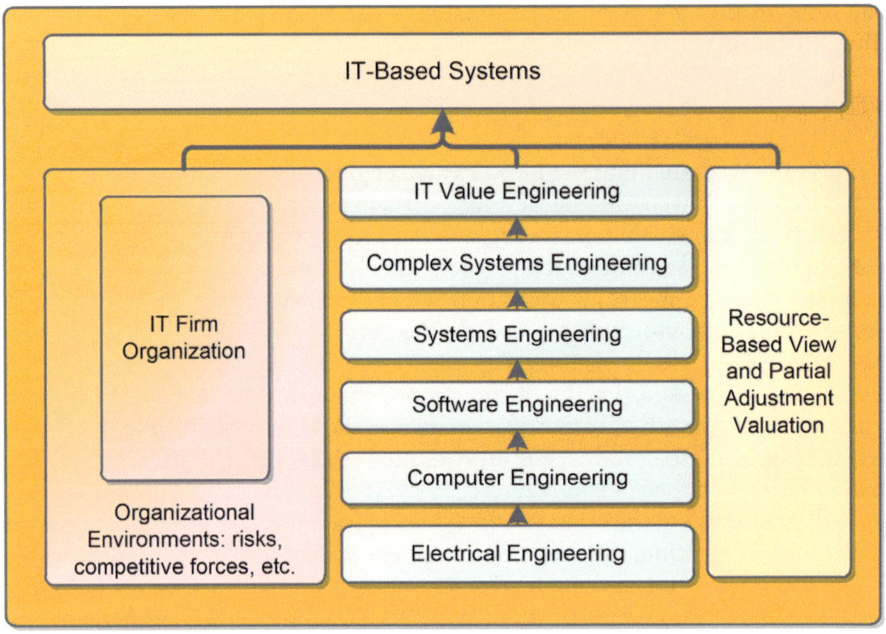

Figure 2.
IT value engineering position in an organizational environment.

knowledge, and maneuverability advantages [5]. In this study, the competitive advantage that becomes a highlight is a cost-competitive advantage, which can result from the IT value engineering through a system optimization effort.

Furthermore, the IT value engineering has presented in an IT-based firm (see **Figure 2**), which is a firm that its core business has two wide-ranging groups of products and services, namely lifespan application development and support services and production processes [5] or industrial products and services that catch, transmit and display data and information by electronic means [49]. In the meantime, using RBV theory, this chapter departs from firm performance towards IT resources instead of the regular RBV, which originates from the resources to the firm performance in terms of the serial configuration.

This chapter proposes to structure the four subsystems of the RBV-based result, namely firm performance (FP), firm competence (FCC), firm capability (FC), and IT resource (ITR) to accomplish the rigorous IT value engineering concept by considering the nature of VRIN (valuable, rare, inimitable, non-substitutable) IT resources. In this case, each subsystem needs to identify its measures, which facilitate to determine the characteristics of the subsystem to build relationships with other subsystems or between the subsystem input and output [50]. Therefore, the FP typically addresses financial and efficiency performance, which manifests in, such as time-to-market and mass customization [51], profitability containing return on investment, return on asset, return on equity [16, 52]. While the FCC emphasizes to a firm's core competence as the learning process ability to manage various resources and technology within the firm [53], consisting of three components: IT knowledge, IT operations, and IT objects. IT knowledge is the extent to which a firm possesses a body of technical knowledge about objects such as computer-based systems, while IT operations are the extent to which a firm utilizes IT to manage market and customer information. IT objects represent computer-based hardware, software, and support personnel [35].

Moreover, FC focuses on the assembling and installing IT-based resource capabilities to work together with the other resources in the firm [11] controlled by IT infrastructures, managed IT skills, and collaboration between IT and business [54]. The three measures combination can result in the firm capability, hence, it can create the VRIN IT resources, which also consist of IT infrastructure as tangible resources, human IT resources representing technical and managerial IT skills, and intangible IT-enabled resources such as knowledge assets, customer orientation, and synergy [11].

Preferably, to construct an IT spending model system, each measure or component of the subsystems relates one to another is not only qualitatively rational, but also quantitatively plausible as issued by [8]. However, to quantitatively plausible, the subsystems should also have complete measures that can manifest in a mathematical model. As proposed above, the mathematical formula to construct the relationship in this research is the partial adjustment valuation theory [20], which involves Cobb–Douglas production transformation as the input function.

4. Partial adjustment valuation (PAV) theory

Dedrick et al. (2003) stated that the failure of the subject area of the relationship between IT spending and the output performance at the firm level occurred due to the difficulty of quantifying measurement between these quantities [8]. Therefore, the chapter tries to do so by PAV. In this case, Nerlove (1958) was a developer of the origin of PAV theory and further developed by the researchers as it is today. The theory tells that the change in real output of a production process generally does not precisely fit the desired output alteration. The alteration measurement is in the present (t), compared with the previous period (t-1) for the real alteration and the desired alteration, which it is clear that there must be a coefficient bridging the relationship between the two alterations called a constant speed of adjustment [19, 20]. Therefore, if written in a mathematical formula, the theory manifests as follows:

$$y_t - y_{t-1} = \mu \left(y_t^* - y_{t-1}\right), \qquad (t = 1, 2, \dots, s) \tag{2}$$

It seems that y_t is the real output of a production process unit, for example, a firm, in time t, as for y_{t-1} is the real output of the equal production process unit at time $t - 1$. While y_t^* is the desired output of the production process unit at time t, and μ is the coefficient depicting a constant speed of adjustment [20]. In an estimation process, an old-fashioned random error symbolized by ϵ_t needs to consider completing the formula. Consequently, Eq. (2) manifests as follows:

$$y_t = \mu y_t^* + (1 - \mu) y_{t-1} + \in_t \qquad (t = 1, 2, \dots, s) \tag{3}$$

Whereas ϵ_t = conventional error. It appears that the real output is equal to the weighted average of the current desired output – with the weights μ – and the real output at a past time, with weights 1-μ. Furthermore, Lin and Kao (2014) suggested that μ in Eq. (2) and (3) can vary and be dynamic, therefore, μ may convert to μ_t where t represents fluctuations in time for the dynamic and μ for the constant or static. This scheme aims to provide more meaning of μ, for instance, the dynamic μ represents the speed of adjustment behavior in connecting the real output alteration with alterations in the desired one. In other words, these two alterations in output comprehend the dynamic nature of μ. Later, the scheme also exhibits the other signification of the state for further exploration [20].

At that time, the writing the equation above can turn to the subsequent Eqs. (4) and (5) [19, 20]:

$$y_t = \mu_t f(X_t;\beta) + (1-\mu_t)y_{t-1} + \in_t \qquad (t = 1,2,\ldots,s) \tag{4}$$

$$\mu_t = g(S_t;\gamma), \quad 0 \leq \mu_t \leq 1, \qquad (t = 1,2,\ldots,s) \tag{5}$$

Here $f(X_t,\beta)$ is the alternate function of the desired output (y_t^*), which manifests in the form of a production function [8, 18, 19, 37, 38]. Accordingly, X_t could consist of a vector of production such as the regular capital (K_t), the regular labor expense (L_t), and the technology spending, in this study related to IT spending (I_t). For the benefit of variable estimation of the production function, it may consist of two compositions. The first is K, L, and I combination to accommodate the factors of capital, labor, and IT spending immediately, and the second is K and L combination that accommodates the factors of capital and labor. Thus, there are two models: $X_t = (K_t, L_t, I_t)$ and $X_t = (K_t, L_t)$ while β is the unknown parameters [19, 20].

Meanwhile, the function $\mu_t = g(S_t; \gamma)$ represents a dynamic speed of adjustment that accommodates variables, which fluctuate along with the different fluctuations of the required output such as return on equity (ROE). The magnitude of μ_t or μ is in the range of 0 and 1 [20], where the value of 0 means that the real output at time t is precisely equal to the real output of the previous period, t-1. While if 1 indicates that the real output is equivalent to the desired output. Conversely, μ_t is a S_t function, a vector of the variable, affecting the speed of adjustment of a firm, and γ is the unknown parameters. Therefore, to return to the original PAV theory, Eq. (4) is as follows:

$$y_t - y_{t-1} = \mu_t f(X_t;\beta) - \mu_t y_{t-1} + \in_t \qquad (t = 1,2,\ldots,s) \tag{6}$$

Essentially the production function of the Eq. (4), namely $f(X_t,\beta)$, can originate from various production functions such as the Cobb–Douglas (CD), the Box–Cox, the Box–Tidwell, the translog, and the constant elasticity of substitution functions [18–20, 37, 38]. The work may select all or a number of them as a test target. For that reason, this study just exploits the CD production function to substitute $f(X_t,\beta)$ in Eq. (4). While, the CD equation is equally in the Eq. (7) below [18]:

$$f(X_t;\beta) = \alpha K_t^{\beta_1} L_t^{\beta_2} I_t^{\beta_3} e^{v_t - u_t} \qquad (t = 1,2,\ldots,s) \tag{7}$$

The Eq. (7) presents the CD function with X_t consisting of production factors K_t, L_t, and I_t. K_t is the regular capital, L_t is the regular labor expense, and I_t is IT capital over time. In other words, Eq. (7) takes into account the IT capital inclusion. Meanwhile α, β_1, β_2, and β_3 are the unknown parameters and $v_t \sim N(0, \sigma_v^2)$, and $u_t \sim |N(0, \sigma_v^2)|$. In addition, to estimate these parameters performs in an estimation process. Equally for the CD function without I_t presence is as follows in Eq. (8):

$$f(X_t;\beta) = \alpha K_t^{\beta_1} L_t^{\beta_2} e^{v_t - u_t} \qquad (t = 1,2,\ldots,s) \tag{8}$$

Justification of the Eq. (8) is equivalent to the Eq. (7), apart just the I_t absence. While the Eq. (5), the speed of adjustment, can display as in Eq. (9) [20]:

$$\mu_t = \gamma_1 + \gamma_2 S_t \qquad \text{with } 0 \leq \mu_t \leq 1 \tag{9}$$

Here μ_t is the dynamic speed of adjustment, and S_t is the dynamic factor that can manipulate the dynamics of μ_t suitable to the time-varying. Likewise, it may show as variances between the actual and the estimable variables of the firm. Furthermore, researchers provide a number of measures to fill these factors with various variables S_t, for example, return on equity, interest rate, firm size, growth option, economic value-added, and Tobin q [19, 20]. While γ_1 and γ_2 are the unknown parameters.

Moreover, if the Eqs. (7) and (9) substitute components of the Eq. (6), it produces an Eq. (10) as follows:

$$\begin{aligned} y_t - y_{t-1} &= \left(\gamma_1 \alpha K_t^{\beta_1} L_t^{\beta_2} I_t^{\beta_3} e^{v_t - u_t}\right) + \left(\gamma_2 S_t \alpha K_t^{\beta_1} L_t^{\beta_2} I_t^{\beta_3} e^{v_t - u_t}\right) \\ &-(\gamma_1 y_{t-1}) - (\gamma_2 S_t y_{t-1}) + \in_t \qquad (t = 1, 2, \dots, s) \end{aligned} \tag{10}$$

The Eq. (10) is for the three-factor production function, namely K_t, L_t, and I_t. It looks that the equation above is analogous to the Eq. (6), except that the production function, namely f (X_t, β), has converted to the Cobb–Douglas function [see Eq. (7)] and the speed of adjustment μ_t replaced by the Eq. (9). The variables and parameters justification of the equation is equivalent to the preceding equations, which substitute it. Meanwhile, for the two-factor function [Eqs. (7) and (8) substituted into the Eq. (6)], the equation becomes Eq. (11) as follows:

$$\begin{aligned} y_t - y_{t-1} &= \left(\gamma_1 \alpha K_t^{\beta_1} L_t^{\beta_2} e^{v_t - u_t}\right) + \left(\gamma_2 S_t \alpha K_t^{\beta_1} L_t^{\beta_2} e^{v_t - u_t}\right) \\ &-(\gamma_1 y_{t-1}) - (\gamma_2 S_t y_{t-1}) + \in_t \qquad (t = 1, 2, \dots, s) \end{aligned} \tag{11}$$

The equation justification is also analogous to the Eq. (10), except just the I_t absence. Furthermore, the Eqs. (10) and (11) are non-linear equations, their solution must also exploit a non-linear least square (NLS) application [20].

5. ITVE methodology

5.1 Meta-analysis

The first method of the ITVE is the meta-analysis approach, where the study concerns with the previous results in the analogous context, namely the relationship between IT resources and business performance. The method enriches the study since various validated hypotheses provide the researcher with strengthening the topic justification, therefore, the study can lead to conclude towards the objective of the chapter qualitatively [13]. In addition, this technique authorizes authors to study several papers addressing the IT value to the business performance relationship from the RBV point of view. Consequently, based on a number of the previous papers, particular topics such as IT resources, firm capabilities, firm core competencies and firm performance are categorically recognizable, where each group has to have relationship one to another for what this relationship leads to a means to link one category to another to construct a model of the IT value. Essentially, the resulting model is not only based on the meta-analysis approach, but also based on the RBV theory.

5.2 Partial adjustment valuation experiment

This method addresses PAV theory, which is linked components of each subsystem to investigate the correlation between IT resources and business performance. This section first reviews the PAV utilization in this chapter through the theory experimentally of the real facts to measure several IT-based firms using the PAV approach to examine the level of the IT value within each firm.

5.2.1 Structure of conceptual models of IT value engineering

According to the meta-analysis, to create conceptual models, which involves logical and mathematical relationships, this method facilitates to develop two types of the model. The first model of IT value or the three-factor model is through substituting the Eq. (6) by the Eq. (7) and (9). Hence, the first model changes to the Eq. (10) above, which is the partial adjustment with Cobb–Douglas (CD) production function is inside and implying K_t (regular capital), L_t (labor expense), and I_t (IT spending) factors.

Meanwhile, through substituting the Eq. (6) by the Eq. (8) and (9) can create the second model of IT value (K_t and L_t) or the two-factor model as comprehended in the Eq. (11) above. Once again, both the Eqs. (10) and (11) are non-linear equations, thus, to estimate them must also use a non-linear least square (NLS) [20].

5.2.2 Determination of the dynamic and static speed of adjustment and the production function

This study selects the return on equity (ROE) [20] as a component of the dynamic factor of the speed of adjustment since the ROE has an adjacent relationship to the regular capital (K), which is the firm equity, thus, the ROE can seem more representative as a dynamic factor (S_t) in this study than the others. Also, in order for the fluctuation rhythm of the K to compensate by the ROE fluctuation. While the ROE is a gain for the year of the parent firm divided by total equity of the parent firm at year-end December. Thus, the ROE becomes a dynamic factor (S_t, t is a period of time) of μ_t (the speed of adjustment) function to signify the dynamics of the speed of adjustment as comprehended in the Eq. (5) or (9). As for the static speed of adjustment, the Eq. (5) or (9) is equal to a constant, which is estimable in the non-linear least squares (NLS) estimation process. Moreover, the production function of the Eq. (4) devotes to the Cobb–Douglas (CD) function as explicated in the Eq. (7) and (8) above due to its simplicity and familiarity in production function transformations [18, 55].

5.2.3 Chapter and estimation models

For the purpose of assessment to separate the presence (with I_{it}) and the absence (without I_{it}) of the IT capital in the PAV approach, the estimate works on both $X_{it} = (K_{it}, L_{it}, I_{it})$ and $X_{it} = (K_{it}, L_{it})$. Here i = 1, ..., r = 8, for example, for the number of testing firms, and t = 1,..., s = 11, for example, for the period of testing data, such as from 2004 to 2014. It is a time-varying, hence, the system is dynamic, therefore, for that reason, the study models can apply both Eqs. (10) and (11), however, caused the equations to overparameterize due to nonlinear, the estimate also needs the nonlinear least squares (NLS) application [20].

5.2.4 Estimation of the dynamic and static factor of the speed of adjustment

The Eqs. (10) and (11) estimation results in the unknown parameters, including γ_1 and γ_2 of the Eq. (9) for the dynamic speed of adjustment, while the static speed of adjustment is constant for all periods t. Therefore, the dynamic speed of adjustment as in the Eq. (9) is estimable to assess the dynamics of the μ_{it}. In addition, due to covering a period of time, the μ_{it} has the average speed of adjustment (ASA) as well. Assume γ_i estimate as $\hat{\gamma}_i$, the average dynamic speed of adjustment (ASA_i) appears as in the Eq. (12) [20].

$$ASA_i = \sum_{t}^{s} \frac{g\left(S_{i(t)}; \hat{\gamma}_i\right)}{s} \qquad (i = 1, \ldots, r \quad and \quad t = 1, \ldots, s) \tag{12}$$

At this point, $g(S_{it}; \hat{\gamma}_i)$ is the S_{it} function, a vector of the variable, affecting the speed of adjustment. In that case, i and t address firm category and period of time. This method allows the study to comprehend the dynamics of the speed of adjustment and the value disparity between the three-factor (K_{it}, L_{it}, I_{it}) and the two-factor (K_{it}, L_{it}) models, which the meaning of the I_{it} factor appears to contribute to the business performance. Meanwhile, the static factor of the speed of adjustment is constant for all periods t, its average is equivalent to its addition, within the periods divided by the amount of the periods.

5.2.5 Valuation of performance measures of the PAV experiment

In order to evaluate the change of the firm performance due to IT spending, Lin and Kao (2014) proposed the performance measures (PM) of the dynamic (μ_{it}) and static (μ_i) speeds of partial adjustment evaluate the performance change of the processing unit tested. This measurement manifests in Eq. (13) below [19, 20]:

$$PM_t = \mu_t f(X_t; \beta) = g(S_t; \gamma) f(X_t; \beta) \tag{13}$$

To estimate the parameter γ and β, both parameters further designated to become $\hat{\gamma}$ and $\hat{\beta}$, thus the Eq. (13) converts to:

$$PV_{it} = \widehat{PM}_{it} = \hat{\mu}_{i(t)} f(X_{it}; \hat{\beta}_i) = g(S_{i(t)}; \hat{\gamma}_i) f(X_{it}; \hat{\beta}_i) \tag{14}$$

In this case, PV_{it} is performance values of the processing unit or the firm. If averaged, the Eq. (14) results in:

$$APVi = \sum_t \frac{PV_{it}}{s} \quad (i = 1, \dots, r \; and \; t = 1, \dots, s) \tag{15}$$

Both the Eqs. (14) and (15) result in the currency value, however, that is further common, it would be superior if presented in the form of an index ratio. Consequently, PV_{it} should be divided by the real output (y_{it}), instead of a "devisor" (y_{it}^{Δ}) as suggested by [19] to become an index of performance ratio (PR). Therefore, the equation seems as the Eq. (6), and if averaged, the equation becomes the Eq. (7), which it can measure to what extent value the role of IT spending in the business organization, compared with no the investment.

$$PR_{it} = \frac{PV_{it}}{y_{it}} \quad (i = 1, \dots, r \; and \; t = 1, \dots, s) \tag{16}$$

The average value (APR) of Eq. (16) appears as using the subsequent formula:

$$APR_i = \sum_t \frac{PR_{it}}{s} \quad (i = 1, \dots, r \; and \; t = 1, \dots, s) \tag{17}$$

Using this method, it is plausible to consider the amount of value between the IT capital presence and its absence within a capital expenditure of the firm. In other words, the IT value model using the PAV guides the study to comprehend the value of IT.

In order to evaluate the change in the firm performance due to IT spending, Lin and Kao (2014) proposed the performance measures (PM) of the dynamic (μ_{it}) and static (μ_i) speeds of partial adjustment to evaluate the performance change of the processing unit tested. This measurement manifests in Eq. (18) below [19, 20]:

$$PM_t = \mu_t f(X_t;\beta) = g(S_t;\gamma)f(X_t;\beta) \tag{18}$$

To estimate the parameter γ and β, both parameters further designated to be $\hat{\gamma}$ and $\hat{\beta}$, thus the Eq. (18) converts to:

$$PV_{it} = \widehat{PM}_{it} = \hat{\mu}_{i(t)} f\left(X_{it};\hat{\beta}_i\right) = g\left(S_{i(t)};\hat{\gamma}_i\right) f\left(X_{it};\hat{\beta}_i\right) \tag{19}$$

In this case, PV_{it} is the performance values of the processing unit or the firm. If averaged, the Eq. (19) results in:

$$APVi = \sum_t \frac{PV_{it}}{s} \quad (i = 1, \dots, r \text{ and } t = 1, \dots, s) \tag{20}$$

Both the Eqs. (19) and (20) result in the currency value, however, that is further common, it would be superior if presented in the form of an index ratio. Consequently, PV_{it} should be divided by the real output (y_{it}), instead of a "devisor" (y_{it}^{Δ}) as suggested by [19] to be an index of performance ratio (PR). Therefore, the equation seems as the Eq. (21), and if averaged, the equation becomes the Eq. (22), which it can measure to what extent value the role of IT spending in the business organization, compared with no the investment.

$$PR_{it} = \frac{PV_{it}}{y_{it}} \quad (i = 1, \dots, r \text{ and } t = 1, \dots, s) \tag{21}$$

The average value (APR) of Eq. (21) is calculated using the subsequent formula:

$$APR_i = \sum_t \frac{PR_{it}}{s} \quad (i = 1, \dots, r \text{ and } t = 1, \dots, s) \tag{22}$$

Using this method, it is plausible to consider the amount of value between the IT capital presence and the absence of it within a capital expenditure of the firm. In other words, the IT value model using the PAV guides the study to comprehend the value of IT.

5.3 The PAV approach validation

In essence, the applied method in this chapter is identical to the abovementioned method, namely starting from the structure of the conceptual model of IT value consisting of two types of models: three and two-factor models up until valuation of performance measures. However, the difference is simply on the goal, namely the earlier method aims to examine the PAV theory using the real facts to make sure that the IT inclusion in the business organization is material and valuable, while, this subchapter is to validate the resulted experiment data in several IT-based firms to certify that the PAV theory encounters the criteria of system measurements from a statistical point of views [8] to identify the level of the IT value of each firm. The validation is through model data examinations.

Here, the exploited data have been covering the period, for example, from 2004 to 2014, collected from the audited financial statements and the published annual reports. To compare between the presence (with I_t) and absence (without I_t) of the IT capital in the PAV approach [20], the estimation involves both $X_t = (K_t, L_t, I_t)$, and $X_t = (K_t, L_t)$ where t = 1..., 11 at the time of confirmed data from 2004 to 2014 for both static and dynamic speed of adjustment.

5.4 Development of IT value engineering model

In reality, the adopted chapter method respects with the exposure of systems engineering processes offered by [50, 56], which is afterwards packaged in the method sequences as depicted below [10].

5.4.1 Definition of the problem

As mentioned, the primary problem of this chapter is how to carry out the need of worthy performance of the IT-based business organization to sustain competitive advantages by optimal costs, especially IT costs. Since this problem involves a variety of factors such as functional subsystems of RBV point of views, financial systems, competitive forces, business performance, risk management, resource management, and so forth. Accordingly, to solve this problem needs a systems engineering approach integrating various components into a unity solving the needed values.

5.4.2 Invention, evaluation, and selection of alternative solutions

In order to solve the problem, various alternative solutions could be a means to undo. Examples of the alternatives are with increasing the firm performance while the IT capital is constant, improving the IT competency and capability of the organization, and cost optimization by encouraging innovation, restructuring, IT cost-saving/ efficiency, and effective IT procurement. Indeed, each alternative has advantages and disadvantages, therefore, the preferred solution is all alternatives combinations to compile in a systems engineering process.

In the meantime, the preferred solution selected based on the five criteria that Kosky et al. (2013) initiated, namely "minimize information content, maintain the independence of functional requirements, ease of manufacture, robustness, and design for adjustability" [10].

5.4.3 Detail of design

According to [56], the systems engineering life cycle phases and the systems engineering method merges, which denotes that for each engineering phase of a horizontal nature, is vertically explored using these engineering models. This step is for concept development and engineering development phases, including each block of the phases. Meanwhile, the post-development phase is beyond this study. Consequently, the analytical results separated into two tables.

5.4.4 Development and validation of the model

Furthermore, the information technology value engineering model exists to develop three types of models: parallel, serial, and hybrid ITVE. Likewise, their validation takes place to certify that the model is reasonable philosophically and technically.

5.4.4.1 Parallel approach model

The parallel model is in **Figure 3** [25, 26]. This figure explicates that the principal subsystems of the model consist of firm performance (FP), firm core competence (FCC), firm capability (FC), and IT resource (ITR), which each subsystem links one to another in a parallel fashion. In a mathematical relationship, the parallel connection manifests an add operation (see **Figure 3**). It implies that the input (y_t^*) is proportionally divided into four sub-inputs, i.e. y_{1t}^*, y_{2t}^*, y_{3t}^*, and y_{4t}^* or

$y_t^* = y_{1t}^* + y_{2t}^* + y_{3t}^* + y_{4t}^*$. Each subsystem has each speed of adjustment (μ_{it}, i = 1,2,3,4 and t = period), i.e. FP has μ_{1t}, FCC has μ_{2t}, FC has μ_{3t}, and ITR has μ_{4t}, whether static (constant) or dynamic [20]. Likewise, the output consists of four sub outputs, i.e. y_{1t}, y_{2t}, y_{3t}, and y_{4t}, which can appear as $y_t = y_{1t} + y_{2t} + y_{3t} + y_{4t}$.

Using the partial adjustment valuation approach [see the Eq. (3)], each subsystem could be mathematically revealed as follows [25, 26], see **Figure 3**:

Firm Performance (FP):

$$y_{1_t} - y_{1_{t-1}} = \mu_1 \left(y_{1_t}^* - y_{1_{t-1}} \right) \tag{23}$$

$$y_{1_t} = \mu_1 y_{1_t}^* + \ (1 - \mu_{1_t}) y_{1_{t-1}} \tag{24}$$

Firm Core Competence (FCC):

$$y_{2_t} - y_{2_{t-1}} = \mu_2 \left(y_{2_t}^* - y_{2_{t-1}} \right) \tag{25}$$

$$y_{2_t} = \mu_2 y_{2_t}^* + (1 - \mu_2) y_{2_{t-1}} \tag{26}$$

Firm Capability (FC):

$$y_{3_t} - y_{3_{t-1}} = \mu_3 \left(y_{3_t}^* - y_{3_{t-1}} \right) \tag{27}$$

$$y_{3_t} = \mu_3 y_{3_t}^* + \ (1 - \mu_3) y_{3_{t-1}} \tag{28}$$

Information Technology Resource (ITR):

$$y_{4_t} - y_{4_{t-1}} = \mu_4 \left(y_{4_t}^* - y_{4_{t-1}} \right) \tag{29}$$

$$y_{4_t} = \mu_4 y_{4_t}^* + (1 - \mu_4) y_{4_{t-1}} \tag{30}$$

If Eq. (24), Eq. (26), Eq. (28), and Eq. (30) are together added would result in Eq. (31) [25, 26]:

$$\begin{aligned} y_t = \mu_1 y_{1_t}^* + \ (1 - \mu_1) y_{1_{t-1}} + \ \mu_2 y_{2_t}^* + (1 - \mu_2) y_{2_{t-1}} + \\ \mu_3 y_{3_t}^* + \ (1 - \mu_3) y_{3_{t-1}} + \mu_4 y_{4_t}^* + (1 - \mu_4) y_{4_{t-1}} \end{aligned} \tag{31}$$

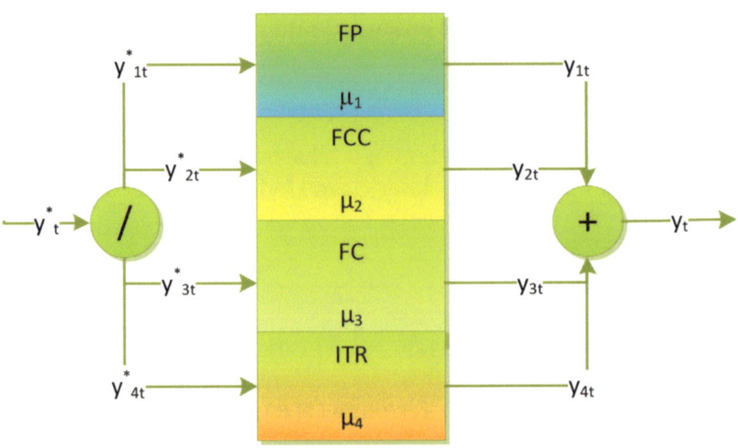

Figure 3.
IT value engineering model in a parallel relationship [25].

Where y_t = the real output of period t, y_{1t} = the real output of FP at period t, y^*_{1t} = the desired output (input) of FP, y_{1t-1} = the real output of the previous period (t-1), and μ_1 = the constant speed of adjustment of FP. Similarly, y_{2t} = the real output of FCC at period t, y^*_{2t} = the desired output (input) of FCC at period t, y_{2t-1} = the real output of the previous period (t-1), and μ_2 = the constant speed of adjustment of FCC. Afterwards, y_{3t} = the real output of FC at period t, y^*_{3t} = the desired output (input) of FC at period t, y_{3t-1} = the real output of the previous period (t-1), and μ_3 = the constant speed of adjustment of FC. Finally, y_{4t} = the real output of ITR at period t, y^*_{4t} = the desired output (input) of ITR period t, y_{4t-1} = the real output of the previous period (t-1), and μ_4 = the constant speed of adjustment of ITR.

5.4.4.2 Serial approach model

Instead of the parallel fashion, the serial ITVEM appears, in which to do so, suppose the Eq. (24), the Eq. (26), the Eq. (28), and the Eq. (30) exhibit in a serial relationship (see **Figure 4**), with an assumption that each output of a subsystem fully becomes an input of the subsequent ones, the end result is as Eq. (32) [25, 26].

$$\begin{aligned} y_t = {} & \left[\mu_1 y_t^* + (1-\mu_1) y_{1_{t-1}}\right] + \left[\mu_2\mu_1 y_t^* + \mu_2(1-\mu_1)y_{1_{t-1}} + (1-\mu_2)y_{2_{t-1}}\right] \\ & + \left[\mu_3\mu_2\mu_1 y_t^* + \mu_3\mu_2(1-\mu_1)y_{1_{t-1}} + \mu_3(1-\mu_2)y_{2_{t-1}} + (1-\mu_3)y_{3_{t-1}}\right] \\ & + \left[\mu_4\mu_3\mu_2\mu_1 y_t^* + \mu_4\mu_3\mu_2(1-\mu_1)y_{1_{t-1}} + \mu_4\mu_3(1-\mu_2)y_{2_{t-1}} + \mu_4(1-\mu_3)y_{3_{t-1}} + (1-\mu_4)y_{t-1}\right] \end{aligned} \tag{32}$$

As for the explanation of the symbols is equal to the parallel ITVE.

5.4.4.3 Hybrid approach model

The hybrid configuration [27] is an option for structuring each subsystem in the chapter. **Figure 5** explicates that the principal subsystems of the model consist of ITR, FC, FCC, and FP. It appears that the resources are the ITR consisting of the regular capital (K_t), the regular labor expense (L_t), and the technology spending, in this chapter related to IT spending (I_t). Furthermore, the resources become inputs of the FC subsystem as K_{cap}, L_{cap}, and I_{cap} to be processed in resulting the FC output, viz. w_{mt} (m = 1,2,3) or w_{1t}, w_{2t}, and w_{3t}, see **Figure 5**. Likewise, the resources also become inputs of the FCC subsystem as K_{com}, L_{com}, and I_{com} to be processed in resulting the FC output, viz. v_{jt} (j = 1,2,3) or v_{1t}, v_{2t}, and v_{3t}, see

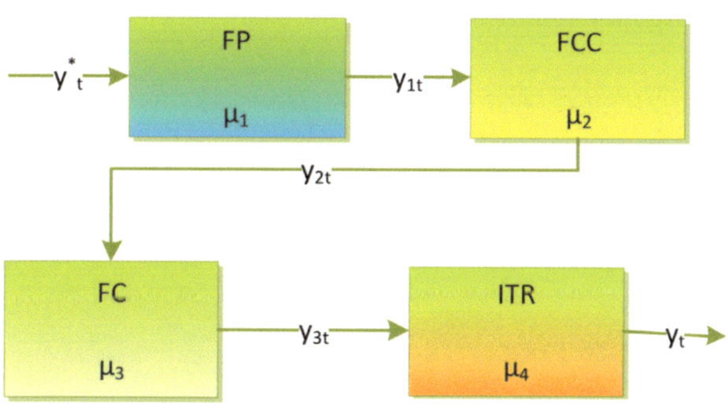

Figure 4.
IT value engineering model in a serial relationship [25].

Figure 5. Moreover, the output of both FC and FCC turn into the input of the FP. In other words, (w_{1t}, w_{2t}, w_{3t}) and (v_{1t},v_{2t}, v_{3t}) appear as inputs of the FP.

Therefore, the PAV model of the hybrid configuration (see **Figure 5**) is as follows [27]:

Firm Capabilities "cap" (FC):

$$w_{m_t} - w_{m_{t-1}} = \mu_m \left(\alpha_m K^{\beta_{1m}}{}_{cap\,m} L^{\beta_{2m}}{}_{cap\,m} I^{\beta_{3m}}{}_{cap\,m} - w_{m_{t-1}}\right) \quad (m = 1, 2, 3; t = 1, \dots, 11) \tag{33}$$

or

$$w_{m_t} = \mu_m \alpha_m K^{\beta_{1m}}{}_{cap\,m} L^{\beta_{2m}}{}_{cap\,m} I^{\beta_{3m}}{}_{cap\,m} + (1 - \mu_m) w_{m_{t-1}} \tag{34}$$

Where w_{mt} = the real output of FC at period t, μ_m = the constant speed of adjustment of FC, α_m = a constant of Cobb–Douglas function; β_1, β_2, and β_3 are input elasticity of production factors regarding the regular capital (K), the labor expense (L), and the IT capital (I), and w_{mt-1} = the real output of the previous period (t-1). Hence, if the FC consists of three variables (m = 1, 2, and 3), viz. IT infrastructures, IT managerial skills, and Collaboration [54], thus each variable has output as follows [27].

IT infrastructures (w_{1t}):

$$w_{1_t} = \mu_1 \alpha_1 K^{\beta_{11}}{}_{cap\,1} L^{\beta_{21}}{}_{cap\,1} I^{\beta_{31}}{}_{cap\,1} + (1 - \mu_1) w_{1_{t-1}} \tag{35}$$

IT managerial skills (w_{2t}):

$$w_{2_t} = \mu_2 \alpha_2 K^{\beta_{12}}{}_{cap\,2} L^{\beta_{22}}{}_{cap\,2} I^{\beta_{32}}{}_{cap\,2} + (1 - \mu_2) w_{2_{t-1}} \tag{36}$$

Collaboration (w_{3t}):

$$w_{3_t} = \mu_3 \alpha_3 K^{\beta_{13}}{}_{cap\,3} L^{\beta_{23}}{}_{cap\,3} I^{\beta_{33}}{}_{cap\,3} + (1 - \mu_3) w_{3_{t-1}} \tag{37}$$

Firm Core Competence "com" (FCC):

$$v_{j_t} - v_{j_{t-1}} = \delta_j \left(\lambda_j K^{\sigma_{1j}}{}_{com\,j} L^{\sigma_{2j}}{}_{com\,j} I^{\sigma_{3j}}{}_{com\,j} - v_{j_{t-1}}\right) \quad (j = 1, 2, 3; t = 1, \dots, 11) \tag{38}$$

or

$$v_{j_t} = \delta_j \lambda_j K^{\sigma_{1j}}{}_{com\,j} L^{\sigma_{2j}}{}_{com\,j} I^{\sigma_{3j}}{}_{com\,j} + (1 - \delta_j) v_{j_{t-1}} \tag{39}$$

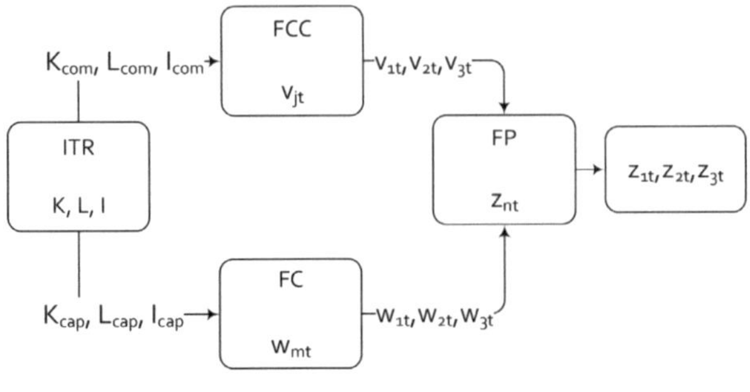

Figure 5.
IT value in the hybrid configuration [27].

Where v_{jt} = the real output of FCC at period t, λ_j = the constant speed of adjustment of FCC, λ_j = a constant of Cobb–Douglas function; σ_1,σ_2, and σ_3 are input elasticity of production factors regarding the regular capital (K), the labor expense (L), and the IT capital (I), and v_{jt-1} = the real output of the previous period (t-1). Hence, if the FCC consists of three variables (j = 1, 2, and 3), viz. IT knowledge, IT operations, and IT objects [35], thus each variable has output as follows [27].

IT knowledge (v_{1t}):

$$v_{1_t} = \delta_1 \lambda_1 K^{\sigma_{11}}{}_{com\,1} L^{\sigma_{21}}{}_{com\,1} I^{\sigma_{31}}{}_{com\ 1} + (1-\delta_1) v_{1_{t-1}} \tag{40}$$

IT managerial skills (v_{2t}):

$$v_{2_t} = \delta_2 \lambda_2 K^{\sigma_{12}}{}_{com\,2} L^{\sigma_{22}}{}_{com\,2} I^{\sigma_{32}}{}_{com\ 2} + (1-\delta_2) v_{2_{t-1}} \tag{41}$$

Collaboration (v_{3t}):

$$v_{3_t} = \delta_3 \lambda_3 K^{\sigma_{13}}{}_{com\,3} L^{\sigma_{23}}{}_{com\,3} I^{\sigma_{33}}{}_{com\ 3} + (1-\delta_3) v_{3_{t-1}} \tag{42}$$

Firm Performance “per” (FP):

$$z_{n_t} - z_{n_{t-1}} = \eta_n \left[\gamma_n (w_{1_t}, w_{2_t}, w_{3_t})^{\varphi_{1n}}{}_{per\,n} (v_{1_t}, v_{2_t}, v_{3_t})^{\varphi_{2n}}{}_{per\,n} - z_{n_{t-1}} \right] \\ (n = 1, 2, 3; t = 1, \ldots, 11) \tag{43}$$

or

$$z_{n_t} = \eta_n \gamma_n (w_{1_t}, w_{2_t}, w_{3_t})^{\varphi_{1n}}{}_{per\,n} (v_{1_t}, v_{2_t}, v_{3_t})^{\varphi_{2n}}{}_{per\,n} + (1-\eta_n) z_{n_{t-1}} \tag{44}$$

Where z_{nt} = the real output of FP at period t, η_n = the constant speed of adjustment of FC, γ_n = a constant of Cobb–Douglas function; ϕ_1 and ϕ_2 are input elasticity of production factors regarding the FC output (w_{1t},w_{2t},w_{3t}) and the FCC output (v_{1t},v_{2t},v_{3t}), and z_{nt-1} = the real output of the previous period (t-1). Hence, if the FP consists of three variables (n = 1, 2, and 3), viz. ROE, ROA, and Revenue [16], thus each variable has output as follows [27].

ROE (z_{1t}):

$$z_{1_t} = \eta_1 \gamma_1 (w_{1_t}, w_{2_t}, w_{3_t})^{\psi_{11}}{}_{per\,1} (v_{1_t}, v_{2_t}, v_{3_t})^{\varphi_{21}}{}_{per\,1} + (1-\eta_1) z_{1_{t-1}} \tag{45}$$

ROA (z_{2t}):

$$z_{2_t} = \eta_2 \gamma_2 (w_{1_t}, w_{2_t}, w_{3_t})^{\varphi_{21}}{}_{per\,2} (v_{1_t}, v_{2_t}, v_{3_t})^{\varphi_{22}}{}_{per\,2} + (1-\eta_2) z_{2_{t-1}} \tag{46}$$

Revenue (z_{3t}):

$$z_{3_t} = \eta_3 \gamma_3 (w_{1_t}, w_{2_t}, w_{3_t})^{\varphi_{31}}{}_{per\,3} (v_{1_t}, v_{2_t}, v_{3_t})^{\varphi_{23}}{}_{per\,3} + (1-\eta_3) z_{3_{t-1}} \tag{47}$$

6. System optimization

The ITVE optimization involves the cost minimization in accordance with the major problem of this research to raise the firm performance at optimal cost [26]. To do so, it needs several assumptions [57] along with the optimization process. For example, the Cobb–Douglas production function [20] replaces each the desired output (the starred y^*_{it}, i = 1, 2, 3, 4 and t = 1, ..., 11, for example) of subsystems. The Cobb-Douglass function is as follows:

$$y_{it}^{*} = \alpha K_{it}^{\beta_1} L_{it}^{\beta_2} I_{it}^{\beta_3} \quad (i = 1, \dots, 4 \ \ and \ \ t = 1, 2, \dots, 11) \tag{48}$$

Whereas y_{it}^{*} = the desired output with i = subsystem and t = period, K_{it} = the regular capital, L_{it} = the labor expense, I_{it} = the IT capital, α = total factor productivity, and β_1, β_2, β_3 = the output elasticity of the regular capital, the labor expense, and the IT capital. Therefore, the partial adjustment for each subsystem is as follows (to simplify, i is disappearing):

$$y_t = \mu_t y_t^{*} + (1 - \mu_t) y_{t-1} = \mu_t \alpha K_t^{\beta_1} L_t^{\beta_2} I_t^{\beta_3} + (1 - \mu_t) y_{t-1} \tag{49}$$

Whereas μ_t is the static speed of adjustment and y_{t-1} is the revenue in the earlier period. Additionally, for cost minimization, the partial derivatives of the Eq. (48) should fulfill these conditions [58, 59]:

$$\frac{\partial y_t}{\partial K_t} = 0, \quad \frac{\partial y_t}{\partial L_t} = 0, \quad \frac{\partial y_t}{\partial I_t} = 0 \tag{50}$$

If the Eq. (49) is mathematically derived to K, L, and I, it respectively results in the following equations (whereas p_1, p_2, and p_3 are added to the equations as unit prices of the regular capital (K), the labor expense (L), and the IT capital (I):

$$\frac{\partial y_t}{\partial K} = \mu_t \alpha \beta_1 p_1 K^{\beta_1 - 1} p_2 L^{\beta_2} p_3 I^{\beta_3} \tag{51}$$

$$\frac{\partial y_t}{\partial L} = \mu_t \alpha \beta_2 p_1 K^{\beta_1} p_2 L^{\beta_2 - 1} p_3 I^{\beta_3} \tag{52}$$

$$\frac{\partial y_t}{\partial I} = \mu_t \alpha \beta_3 p_1 K^{\beta_1} p_2 L^{\beta_2} p_3 I^{\beta_3 - 1} \tag{53}$$

Using the Eq. (50) prerequisites, the Eq. (51) = the Eq. (52) = the Eq. (53), further equations arise as follows:

$$K = \frac{p_3}{p_1} \frac{\beta_1}{\beta_3} I; \quad L = \frac{p_1}{p_2} \frac{\beta_2}{\beta_1} K; \quad and \ I = \frac{p_2}{p_3} \frac{\beta_3}{\beta_2} L \tag{54}$$

If the Eq. (49) is substituted by the Eq. (54) such that the new equation appears in the regular capital (K) variable, the equation is as Eq. (55) and afterwards simplified to become Eq. (56).

$$y_t = \mu_t \alpha K^{\beta_1} \left[\frac{p_1}{p_2} \frac{\beta_2}{\beta_1} K \right]^{\beta_2} \left[\frac{p_1}{p_3} \frac{\beta_3}{\beta_1} K \right]^{\beta_3} + (1 - \mu_t) y_{t-1} \tag{55}$$

$$y_t = \mu_t \alpha \beta_1^{-\beta_2 - \beta_3} \beta_2^{\beta_2} \beta_3^{\beta_3} p_1^{\beta_2 + \beta_3} p_2^{-\beta_2} p_3^{-\beta_3} K^{\beta_1 + \beta_2 + \beta_3} + (1 - \mu_t) y_{t-1} \tag{56}$$

Furthermore, the Eq. (56) becomes K variable as in Eq. (57) and afterwards simplified as in Eq. (58) as follows:

$$K^{\beta_1 + \beta_2 + \beta_3} = \mu_t^{-1} \alpha^{-1} \beta_1^{\beta_2 + \beta_3} \beta_2^{-\beta_2} \beta_3^{-\beta_3} p_1^{-\beta_2 - \beta_3} p_2^{\beta_2} p_3^{\beta_3} \left[y_t - (1 - \mu_t) y_{t-1} \right] \tag{57}$$

$$K = \mu_t^{\frac{-1}{(\beta_1 + \beta_2 + \beta_3)}} \alpha^{\frac{-1}{(\beta_1 + \beta_2 + \beta_3)}} \beta_1^{\frac{\beta_2 + \beta_3}{(\beta_1 + \beta_2 + \beta_3)}} \beta_2^{\frac{-\beta_2}{(\beta_1 + \beta_2 + \beta_3)}} \beta_3^{\frac{-\beta_3}{(\beta_1 + \beta_2 + \beta_3)}} p_1^{\frac{-\beta_2 - \beta_3}{(\beta_1 + \beta_2 + \beta_3)}} p_2^{\frac{\beta_2}{(\beta_1 + \beta_2 + \beta_3)}} p_3^{\frac{\beta_3}{(\beta_1 + \beta_2 + \beta_3)}} \left[y_t - (1 - \mu_t) y_{t-1} \right]^{\frac{1}{(\beta_1 + \beta_2 + \beta_3)}} \tag{58}$$

Using the equivalent way, the variable L and I can become as follows:

$$L = \mu_t^{\frac{-1}{(\beta_1+\beta_2+\beta_3)}} \alpha^{\frac{-1}{(\beta_1+\beta_2+\beta_3)}} \beta_1^{\frac{-\beta_1}{(\beta_1+\beta_2+\beta_3)}} \beta_2^{\frac{\beta_1+\beta_3}{(\beta_1+\beta_2+\beta_3)}} \beta_3^{\frac{-\beta_3}{(\beta_1+\beta_2+\beta_3)}} p_1^{\frac{\beta_1}{(\beta_1+\beta_2+\beta_3)}} p_2^{\frac{-\beta_1-\beta_3}{(\beta_1+\beta_2+\beta_3)}} p_3^{\frac{\beta_3}{(\beta_1+\beta_2+\beta_3)}} [y_t - (1-\mu_t)y_{t-1}]^{\frac{1}{(\beta_1+\beta_2+\beta_3)}} \tag{59}$$

$$I = \mu_t^{\frac{-1}{(\beta_1+\beta_2+\beta_3)}} \alpha^{\frac{-1}{(\beta_1+\beta_2+\beta_3)}} \beta_1^{\frac{-\beta_1}{(\beta_1+\beta_2+\beta_3)}} \beta_2^{\frac{-\beta_2}{(\beta_1+\beta_2+\beta_3)}} \beta_3^{\frac{\beta_1+\beta_2}{(\beta_1+\beta_2+\beta_3)}} p_1^{\frac{\beta_1}{(\beta_1+\beta_2+\beta_3)}} p_2^{\frac{\beta_2}{(\beta_1+\beta_2+\beta_3)}} p_3^{\frac{-\beta_1-\beta_2}{(\beta_1+\beta_2+\beta_3)}} [y_t - (1-\mu_t)y_{t-1}]^{\frac{1}{(\beta_1+\beta_2+\beta_3)}} \tag{60}$$

If K, L, and I are multiplying p_1, p_2, and p_3 as unit prices respectively, then it appears as follows:

$$p_1K = \mu_t^{\frac{-1}{(\beta_1+\beta_2+\beta_3)}} \alpha^{\frac{-1}{(\beta_1+\beta_2+\beta_3)}} \beta_1^{\frac{\beta_2+\beta_3}{(\beta_1+\beta_2+\beta_3)}} \beta_2^{\frac{-\beta_2}{(\beta_1+\beta_2+\beta_3)}} \beta_3^{\frac{-\beta_3}{(\beta_1+\beta_2+\beta_3)}} p_1^{\frac{\beta_1}{(\beta_1+\beta_2+\beta_3)}} p_2^{\frac{\beta_2}{(\beta_1+\beta_2+\beta_3)}} p_3^{\frac{\beta_3}{(\beta_1+\beta_2+\beta_3)}} [y_t - (1-\mu_t)y_{t-1}]^{\frac{1}{(\beta_1+\beta_2+\beta_3)}} \tag{61}$$

$$p_2L = \mu_t^{\frac{-1}{(\beta_1+\beta_2+\beta_3)}} \alpha^{\frac{-1}{(\beta_1+\beta_2+\beta_3)}} \beta_1^{\frac{-\beta_1}{(\beta_1+\beta_2+\beta_3)}} \beta_2^{\frac{\beta_1+\beta_3}{(\beta_1+\beta_2+\beta_3)}} \beta_3^{\frac{-\beta_3}{(\beta_1+\beta_2+\beta_3)}} p_1^{\frac{\beta_1}{(\beta_1+\beta_2+\beta_3)}} p_2^{\frac{\beta_2}{(\beta_1+\beta_2+\beta_3)}} p_3^{\frac{\beta_3}{(\beta_1+\beta_2+\beta_3)}} [y_t - (1-\mu_t)y_{t-1}]^{\frac{1}{(\beta_1+\beta_2+\beta_3)}} \tag{62}$$

$$p_3I = \mu_t^{\frac{-1}{(\beta_1+\beta_2+\beta_3)}} \alpha^{\frac{-1}{(\beta_1+\beta_2+\beta_3)}} \beta_1^{\frac{-\beta_1}{(\beta_1+\beta_2+\beta_3)}} \beta_2^{\frac{-\beta_2}{(\beta_1+\beta_2+\beta_3)}} \beta_3^{\frac{\beta_1+\beta_2}{(\beta_1+\beta_2+\beta_3)}} p_1^{\frac{\beta_1}{(\beta_1+\beta_2+\beta_3)}} p_2^{\frac{\beta_2}{(\beta_1+\beta_2+\beta_3)}} p_3^{\frac{\beta_3}{(\beta_1+\beta_2+\beta_3)}} [y_t - (1-\mu_t)y_{t-1}]^{\frac{1}{(\beta_1+\beta_2+\beta_3)}} \tag{63}$$

Moreover, the Eqs. (61), (62), and (63) substituted into Eq. (64), the total cost of yielding y units in the low-cost technique manifest as the Eq. (64) and (65).

$$\begin{aligned} C(p_1,p_2,p_3,y_t) &= p_1K + p_2L + p_3I \\ &= B\, p_1^{\frac{\beta_1}{(\beta_1+\beta_2+\beta_3)}} p_2^{\frac{\beta_2}{(\beta_1+\beta_2+\beta_3)}} p_3^{\frac{\beta_3}{(\beta_1+\beta_2+\beta_3)}} [y_t - (1-\mu_t)y_{t-1}]^{\frac{1}{(\beta_1+\beta_2+\beta_3)}} \end{aligned} \tag{64}$$

Where B:

$$B = \mu_t^{\frac{-1}{(\beta_1+\beta_2+\beta_3)}} \alpha^{\frac{-1}{(\beta_1+\beta_2+\beta_3)}} \left[\left(\frac{\beta_1}{\beta_2}\right)^{\frac{\beta_2}{(\beta_1+\beta_2+\beta_3)}} \left(\frac{\beta_1}{\beta_3}\right)^{\frac{\beta_3}{(\beta_1+\beta_2+\beta_3)}} + \left(\frac{\beta_2}{\beta_1}\right)^{\frac{\beta_1}{(\beta_1+\beta_2+\beta_3)}} \left(\frac{\beta_2}{\beta_3}\right)^{\frac{\beta_3}{(\beta_1+\beta_2+\beta_3)}} + \left(\frac{\beta_3}{\beta_1}\right)^{\frac{\beta_1}{(\beta_1+\beta_2+\beta_3)}} \left(\frac{\beta_3}{\beta_2}\right)^{\frac{\beta_2}{(\beta_1+\beta_2+\beta_3)}} \right] \tag{65}$$

Whereas p_1, p_2, and p_3 is unit prices of the regular capital (K_t), the labor expense (L_t), and the IT capital (I_t) respectively, y_t is the real output of period t, y_{t-1} is the real output of earlier period t-1, and C is the total cost [26].

7. Conclusion

The significant problem surrounding this study is to sustain superior firm performance as desired at optimal costs due to the IT presence, which has inevitably become a need for running the business world. Numerous studies on the relationship of the firm performance of the IT resource were more focused on a statistical method that links between components using survey data. In essence, this study undertakes an analogous study, but with a different approach, namely, the systems engineering approach combined with RBV theory, systems engineering, the theory of partial adjustment, including the CD production function, which, in turn, lead to creating the ITVE. Furthermore, to create the ITVE, the followed stages are to build the conceptual model of the IT value based on the RBV theory, model experiment using PAV, validate PAV, model the ITVE, confirm the ITVE and study managerial impacts of the model.

The conceptual model of IT value has logically exemplified the relationship between ITR, FC, FCC, and FP in terms of competitive advantages. The theory of partial adjustment links logically the model, which formulates it in two types of models. Explicitly, the first model addresses PAV with the IT capital presence (with I_t) inside of its production function, and the second model with the IT capital absence (without I_t). The applied production function is the CD function while the dynamic factor component of the speed of adjustment is the ROE. However, it may be replaced by other dynamic factors.

The principal problem of this chapter is how to achieve the optimal resources, for instance, IT resource costs, for required business performance. By benefiting the earlier studies, namely the systems engineering methodology, the conceptual model of IT value, the RBV theory, and the PAV theory can solve this problem so that the solution results in the IT value engineering. Furthermore, using the analysis results, a synthesis work leads to composing a block diagram, which depicts a model in terms of the systems engineering of IT value engineering framework, which ultimately results in serial, parallel, and hybrid configurations. Likewise, by benefiting CD production function involved within PAV, the optimal cost of the required firm performance occurs. For that reason, it should surely be an experiment as a simulation on work mechanisms of the model. Consequently, the ITVE technically appears as a framework to study IT value models. However, in practice, this model contributes to managerial implications, which should reinforce the match between techniques and practices.

Author details

Lukman Abdurrahman
Information Systems Program Study, School of Industrial Engineering, Telkom University, Bandung, Indonesia

*Address all correspondence to: abdural@telkomuniversity.ac.id

References

[1] Akkermans H. Intelligent e-business: from technology to value. IEEE INTELLIGENT SYSTEMS. 2001:8-10. DOI: 10.1109/5254.941352

[2] Hadaya, P., and Cassivi, L. Joint collaborative planning as a governance mechanism to strengthen the chain of IT value co-creation. The Journal of Strategic Information Systems. 21. 2012: 182–200.

[3] Kohli, R., Devaraj, S. and Ow, T.T. Does information technology investment influence a firm's market value? A case of non-publicly traded healthcare firms. MIS Quarterly. 36(4). 2012: 1145–1163. DOI:10.2307/41703502

[4] Iansiti, M. And Richards, G.L. The information technology ecosystem: structure, health, and performance. Antitrust Bulletin. 51(1). 2006: 77–110.

[5] Boar, B.H. The art of strategic planning for information technology. 2nd ed. John Wiley and Sons, Inc., New York, USA; 2001.

[6] DeMaagd, K.A. Identifying and measuring the value of information technology in developing economies. [Ph.D. dissertation]. Business Administration: The University of Michigan; 2006.

[7] Dehning, B., Richardson, V.J and Stratopoulos, T. Information technology investments and firm value. Information & Management. 429(7). 2008: 989–1008. DOI: 10.1016/j.im.2004.11.003.

[8] Dedrick, J., Gurbaxani, V. and Kraemer, K.L. Information technology and economic performance: a critical review of the empirical evidence. ACM Computing Surveys. 35(1). March, 2003: 1–28. DOI:10.1145/641865.641866.

[9] Ramos, A.L., Ferreira, J.V., & Barcel'o, J. Model-based systems engineering: an emerging approach for modern systems. IEEE Transaction on Systems, Man, and Cubernetics-Part C: Applications and Reviews. 42(1). 2012: 101–111. DOI: 10.1109/TSMCC.2011.2106495.

[10] Kosky, P., Balmer, R., Keat, W. and Wise, G. Exploring engineering: an introduction to engineering and design. 3rd ed. Elsevier, Waltham, MA, USA; 2013.

[11] Bharadwaj, A.S. A resource-based perspective on information technology capability and firm performance: an empirical investigation. *MIS Quarterl.* 24. 2000: 169–196. DOI: 10.2307/3250983

[12] Ravichandran, T. and Lertwongsatien, C. Effect of information systems resources and capabilities on firm performance: a resource-based perspective. Journal of Management Information Systems. 21. 2005: 237–276.

[13] Liang, T.P., You, J.J. and Liu, C.C. A resource-based perspective on information technology and firm performance: a meta-analysis. Industrial Management & Data Systems. 110(8). 2010: 1138–1158. DOI: 10.1108/02635571011077807.

[14] Muhanna, W.A. and Stoel, M.D. How do investors value IT? An empirical investigation of the value relevance of IT capability and IT spending across industries. Journal of Information Systems. 24(1). 2010: 43–66. DOI: 10.2308/jis.2010.24.1.43

[15] Byrd, T.A., Pitts, J.P., Adrian, A.M. and Davidson, N.W. Examination of a path model relating information technology infrastructure with firm performance. Journal of Business

Logistics. 29(2). 2008: 161–187. DOI: 10.1002/j.2158-1592.2008.tb00091.x

[16] Lin, B.W. Information technology capability and value creation: evidence from the US banking industry. Technology in Society. 29. 2007: 93–106. DOI: 10.1016/j.techsoc.2006.10.003

[17] Kohli, R. and Devaraj, S. Measuring information technology payoff: a meta-analysis of structural variables in firm-level empirical research. Information Systems Research. 14(2). 2003: 127–145. DOI: 10.1287/isre.14.2.127.16019.

[18] Lin, W.T. The business value of information technology as measured by technical efficiency: evidence from country-level data. Decision Support Systems. 46(4). 2009: 865–874. DOI: 10.1016/j.dss.2008.11.017

[19] Lin, W.T., Chuang, C.H. and Choi, J. H. A partial adjustment approach to evaluating and measuring the business value of information technology. Int. J. Production Economics. 127(1). 2010: 158–172. DOI: 10.1016/j.ijpe.2010.05.007.

[20] Lin, W.T. and Kao, T.W. The partial adjustment valuation approach with dynamic and variable speeds of adjustment to evaluating and measuring the business value information technology. European Journal of Operational Research. 238(1). 201: 208–220. DOI: 10.1016/j.ejor.2014.03.019.

[21] Abdurrahman, L., Langi, A.Z.R. and Suhardi. Development of methodologies for measuring IT capability in the information and communication industries. In: IEEE Proceeding of the International Conference on Information Technology Systems and Innovation (ICITSI); 24–27 November 2014, Bandung-Bali: IEEE; 2014. p. 189–194.

[22] Abdurrahman, L., Suhardi and Langi, A.Z.R. Information technology (IT) value model using variance-based structural equation modeling: towards IT value engineering. In: IEEE Proceeding of the Second International Conference on Information and Communication Technology; 28–30 May 2014, Bandung: IEEE; 2014. p. 499–504.

[23] Abdurrahman, L., Suhardi and Langi, A.Z.R. Engineering information technology value in IT-based industries using partial adjustment valuation and resource-based view approach. Int. J. Information and Communication Technology. 8 (4). 2016: 420–435. DOI: 10.1504/IJICT.2016.076824

[24] Abdurrahman, L., Suhardi, Langi, A.Z.R. Valuation methodology of information technology (IT) value in the IT-based business: A case study at a leading telecommunication company. International Journal on Electrical Engineering and Informatics. 8(4). 2016: 864–884. DOI: 10.15676/ijeei.2016.8.4.12.

[25] Abdurrahman, L., Langi, A.Z.R., Suhardi, and Simatupang, T.M. Information technology value engineering model and cost efficiency in IT-based firms. IEEE Systems Journal. 12(3). 2018: 2925–2936. DOI: 10.1109/JSYST.2017.2663418.

[26] Abdurrahman, L., Suhardi, Simatupang, T.M., and Langi, A.Z.R. Information technology value model and its optimal application in IT-based firms. Int. J. Mathematical Modelling and Numerical Optimisation. 8(4). 2018: 331–350. DOI: 10.1504/IJMMNO.2018.10015800.

[27] Abdurrahman, L. Hybrid Configuration in Information Technology Value Model. IEEE Systems Journal. 14(4). 2020: 4515–4522. DOI: 10.1109/JSYST.2020.2968415.

[28] Kohli, R. and Devaraj, S. Realizing the business value of information

technology investments: an organizational process. MIS Quarterly Executive. 3(1). 2004: 53–68.

[29] Gidumal, J.B. and González, M. Maximizing the positive influence of IT for improving organizational performance. Journal of Strategic Information Systems. 20. 2011: 461–478. DOI: 10.1016/j.jsis.2011.09.004.

[30] Mithas, S., Tafti, A., Bardhan, I. and Goh, J.M. Information technology and firm profitability: mechanisms and empirical evidence. MIS Quarterly. 36 (1). 2012: 205–224. DOI: 10.2307/41410414.

[31] Piatkowski, M. Does ICT investment matter for growth and labor productivity in transition economies [Internet]. 2005. Available from http://ssrn.com/abstract=510082.

[32] Kim, G., Shin, B., Kim, K.K, and Lee, H.G. IT capabilities, process-oriented dynamic capabilities, and firm financial performance. Journal of the Association for Information Systems. 12. 2011: 487–517. DOI: 10.17705/1jais.00270.

[33] Cao, G., Wiengarten, F., and Humphreys, P. Towards a contingency resource-based view of IT business value. Syst Pract Action Res. 2011: 85–106. DOI: 10.1007/s11213-010-9178-0.

[34] Kohli, R. and Grover,V. Business value of IT: an essay on expanding research directions to keep up with the times. Journal of the Association for Information Systems. 9 (1). 2008: 23–39. DOI: 10.17705/1jais.00147.

[35] Tippins, M.J., and Sohi, R.S. IT competency and firm performance: Is organizational learning a missing link? Strategic Management Journal. 24. 2003: 745–761. DOI: 10.1002/smj.337.

[36] Mithas, S., Ramasubbu, N. and Sambamurthi, V. How information management capability influences firm performance. MIS Quarterly. 35(1). 2011: 237–256. DOI: 10.2307/23043496.

[37] Lin, W.T. and Chiang, C.Y. The impacts of country characteristics upon the value of information technology as measured by productive efficiency. Int. J. Production Economics. 132(1). 2011: 13–33.

[38] Lin, W.T. and Chuang, C.H. Investigating and comparing the dynamic patterns of the business value of information technology over time. European Journal of Operational Research. 228(1). 2013: 249–261. DOI: 10.1016/j.ejor.2013.01.023.

[39] Goh, K.H. Process and outcome-related IT value: innovations in theory and methods for measurement. [Ph.D. dissertation]. Minneapolis: Faculty of The Graduate School, University of Minnesota; 2007.

[40] Rau S.E. and Bye, B.S. Are you getting value from your IT? Journal of Business Strategy. 2003: 16–20.

[41] Wagner, H.T. and Weitzel, T. Towards an IT production function: understanding routines as fundamental for IT value creation. Journal of Enterprise Information Management. 20(4). 2007: 380–395.

[42] Webb, A. Value engineering part 1. Engineering Management Journal. 1993: 171–175.

[43] Webb, A. Value engineering part 2. Engineering Management Journal. 1993: 231–235.

[44] Jin-wen, Z. and Xiao-ying, Z. The new tools of enterprise's innovation management: value engineering. In: Proceeding of International Conference on Information Management, Innovation Management and Industrial Engineering; 2009. p. 507–510.

[45] Tse, D.C. A conceptual model for assessing the value of information technology [Master thesis]. Ontario, Canada: Management Sciences, University of Waterloo; 2006.

[46] Pang, M.S. *Information technology and value creation in the public sector organizations* [Ph.D. dissertation]. USA: Business Administration, The University of Michigan; 2011.

[47] Tohidi, H. Review the benefits of using value engineering in information technology project management. Procedia Computer Science. 3. 2011: 917–924. DOI: 10.1016/j.procs.2010.12.150.

[48] Sharma A., and Belokar, R.M. Achieving success through value engineering: a case study. In: Proceedings of the World Congress on Engineering and Computer Science; 2012.

[49] International standard industrial classification of all economic activities. 2015. Availabble from http://unstats.un.org/unsd/statcom/doc02/isic.pdf. [Accessed: January 8, 2015].

[50] Lightfoot, R.S. Enhanced concurrent engineering: an approach for accelerating the systems engineering process while minimizing the risk. In: IEEE International Engineering Management Conference 2. IEEE; 2002. p . 779–782.

[51] Byrd, T.A. Information technology, core competencies, and sustained competitive advantage. Information Resources Management Journal. 2001: 27–36. DOI: 10.4018/irmj.2001040103.

[52] Cardeal, N. and António, N. Valuable, rare, inimitable resources and organization (VRIO) resources or valuable, rare, inimitable resources (VRI) capabilities: what leads to competitive advantage? African Journal of Business Management. 6(37). 2012: 10159–10170. DOI: 10.5897/AJBM12.295.

[53] Prahalad, C.K. and Hamel, G. The core competence of the corporation. Harvard Business Review. 1990: 79–90.

[54] Yin, G. and Yang, B. The construction of firm's IT capability and its impact on IT assimilation: an empirical investigation in China. In: Pasific Asia Conference on Information Systems; 9–12 July 2010; Taipei, Taiwan: 2010. p . 11–20.

[55] Hong, G.E., Yu-wei, GUO. and Ming-di, LI. Measurement of IT returns on firm with multiple production processes by simultaneous production functions. In: *Proceeding of the International Conference on Management Science & Engineering (20th)*, July 17–19; Harbin, China: 2013.

[56] Kossiakoff, A., Sweet, W.N, Seymour, S. J. and Biemer, S.M. Systems engineering: principles and practice. John Wiley and Sons, Inc., Hoboken, New Jersey; 2011.

[57] Townsend, J.T. Serial vs parallel processing: sometimes they look like tweedledum and tweedledee but they can (and should) be distinguish. Psychological Science. 1(1). 1990: 46–54.

[58] Hu, Z.H. Reliable optimal production control with cobb-douglas model. Reliable Computing. 4. 1998: 63–69.

[59] Fraser, I. The cobb-douglas production function: an antipodean defence? Economic Issues. 7(1). 2002.

Chapter 13

Multi-Agent Implementation of Filtering Multiset Grammars

Igor Sheremet

Abstract

Chapter is dedicated to the application of multi-agent technology to generation of sets of terminal multisets (*TMS*) defined by filtering multiset grammars (*FMG*). Proposed approach is based on creation of multi-agent system (*MAS*), corresponding to specific *FMG* in such a way, that every rule of *FMG* is represented by independently acting agent. Such *MAS* provides high-parallel generation of *TMS* and may be effectively used in any proper hardware environment. Directions of further development of the proposed approach are discussed.

Keywords: multi-agent systems and technologies, multisets, multiset grammars, filtering multiset grammars, parallel computations

1. Introduction

Filtering multiset grammars (*FMG*) were developed as a result of multiset-based deep integration and convergence of classical mathematical programming and modern knowledge engineering for compact, flexible and natural representation of wide spectrum of combinatorial problems and their solution by application of unified algorithmics [1–6].

One of the advantages of *FMG* is natural parallelism of generation of multisets (*MS*) due to the possibility of independent application of rules to the currently available *MS*. Such feature being supported by appropriate hardware is a very promising background for the effective implementation of *FMG* and hence effective solution of the aforementioned problems. However, the "brutal force" approach, demanding on the extensive parallelism, is not suitable here because of evident cost restrictions. More attractive looks such techniques which would be based on the "branches and bounds" logics providing cut off sets of multisets (*SMS*) which provably do not contain terminal multisets (*TMS*), defined by *FMG*, without their generation.

To develop such perspective approach we propose to apply multi-agent technology (*MAT*) [7–12] as a basis for implementation of the aforementioned techniques. The main idea of the suggested method of *TMS* generation is representation of the generating engine as a multi-agent system (*MAS*), which includes:

N agents, each corresponding to one rule from *FMG* scheme;
one agent corresponding to *FMG* filter;
one supervising agent, accumulating generated *TMS*, satisfying filter.

This *MAS* operates in such a way that every rule is applied (i.e. agent becomes active) as soon as there occurs multiset, matching this rule. By this approach maximally possible degree of parallelism is achieved.

Structure of the chapter is as follows. Section 2 contains main definitions and notions of the multigrammatical framework, necessary for further considerations. Proposed techniques of the multi-agent implementation of multisets generation is introduced in Section 3. Directions of future development of these techniques are discussed in the conclusion.

2. Basic notions and definitions of the multigrammatical framework

Classical theory of sets is based on notion of set as an unordered collection of mutually distinguishable elements. Basic assumption of theory of multisets is that aforementioned collection may contain indistinguishable (identical) elements:

$$v = \left\{ \underbrace{a_1, \ldots, a_1}_{n_1 \ times}, \ldots, \underbrace{a_i, \ldots, a_i}_{n_i \ times}, \ldots, \underbrace{a_m, \ldots, a_m}_{n_m \ times} \right\}. \quad (1)$$

Record (1) is usually represented as

$$v = \{n_1 \cdot a_1, \ldots, n_m \cdot a_m\}, \quad (2)$$

where v is called *multiset*, $n_i \cdot a_i$ – *multiobjects* (*MO*), a_i – *objects*, n_i– their *multiplicities*, for all $i = 1, \ldots, m$. According to (2), multiset may be considered as a set of multiobjects, and, in fact, multiset $\{1 \cdot a_1, \ldots, 1 \cdot a_m\}$ and set $\{a_1, \ldots, a_m\}$ represent one and the same collection. Set$\{a_1, \ldots, a_m\}$, denoted as $\beta(v)$, is called *basis* of multiset v. Both empty multiset and empty set are denoted as $\{\emptyset\}$.

Zero multiplicity of some object is equivalent to the absence of this object in multiset, i.e.

$$\{n_1 \cdot a_1, \ldots, n_m \cdot a_m, 0 \cdot a_{m+1}\} = \{n_1 \cdot a_1, \ldots, n_m \cdot a_m\}. \quad (3)$$

Fact, that object a *enters MS* v (or *MS* v *includes* object a), is denoted as $a \in v$. Symbol "$\in$" is also used to denote, that *MO* $n \cdot a$ enters *MS* v (*MSv* includes *MO* $n \cdot a$):$n \cdot a \in v$. Structure of the left operand determines what kind of relation is referred in every particular case. Similarly, symbol "$\notin$" is used to denote, that object a (multiobject $n \cdot a$) does not enter multiset v. Due to (3), $a \notin v$ and $0 \cdot a \in v$ are equivalent.

Number $|v| = m$ is called *dimensionality* of *MS* v, and number

$$|v| = \sum_{i=1}^{m} n_i, \quad (4)$$

is called *power* of *MS* v.

Two basic relations on multisets – *inclusion* ("$\subseteq$") and *strict inclusion* ("$\subset$") – are defined as follows.

MS v is *included* to *MS* v', if

$$(\forall n \cdot a \in v)(\exists n' \cdot a \in v')\, n \leq n', \quad (5)$$

i.e. for every *MO* $n \cdot a$ entering *MS* v there exists *MO*$n' \cdot a$, which multiplicity n' is not less than n. There may be also $n' \cdot a' \in v'$ such that $a' \notin v$ (as seen, this does not contradict (5), because$0 \cdot a' \in v$ and $0 < n'$).

If $v \subseteq v'$ and $v \neq v'$, then *MS* v is *strictly included* to *MS* v', that is denoted $v \subset v'$.

MS v, which is included to *MS* v', is called *submultiset* of *MS* v'; *MS* v, which is strictly included to *MS* v', is called *strict submultiset* of *MS* v'.

Let us illustrate introduced notions by the following example, where, as in all other examples, having place in this chapter, objects will be represented by strings in brackets.

Example 1. Let

$$v = \{1 \cdot (eur), 5 \cdot (usd), 12 \cdot (rur)\},$$
$$v' = \{6 \cdot (eur), 5 \cdot (usd), 12 \cdot (rur)\}.$$

As seen, according to (5), $v \subseteq v'$ and $v \subset v'$.∎

Three basic operations on multisets are *multiplication by a constant*, *addition*, and *subtraction*, which are denoted by bold symbols $*$, $+$ and $-$ respectively. Semantics of these operations is defined by use of the well known set-theoretical operations (join and intersection), as well as arithmetic operations on integer numbers:

$$n * \{n_1 \cdot a_1, \dots, n_m \cdot a_m\} = \{(n \times n_1) \cdot a_1, \dots, (n \times n_m) \cdot a_m\}, \tag{6}$$

where "$\times$" denotes multiplication of integer numbers;

$$v + v' = \bigcup_{\substack{a \in \beta(v) \cup \beta(v') \\ n \cdot a \in v \\ n' \cdot a \in v'}} \{(n + n') \cdot a\}, \tag{7}$$

$$v - v' = \bigcup_{\substack{a \in \beta(v) \cup \beta(v') \\ n \cdot a \in v \\ n' \cdot a \in v' \\ n > n'}} \{(n - n') \cdot a\}. \tag{8}$$

Along with these operations, we shall use set-theoretical operations on multisets – *join* and *intersection*, – denoted respectively by bold symbols $\cup$ and $\cap$, different from $\cup$ and $\cap$:

$$v \cup v' = \bigcup_{\substack{a \in \beta(v) \cup \beta(v') \\ n \cdot a \in v \\ n' \cdot a \in v'}} \{\max\{n, n'\} \cdot a\}, \tag{9}$$

$$v \cap v' = \bigcup_{\substack{a \in \beta(v) \cup \beta(v') \\ n \cdot a \in v \\ n' \cdot a \in v'}} \{\min\{n, n'\} \cdot a\}, \tag{10}$$

We have used equivalence between $a \notin v$ and $0 \cdot a \in v$ in (9).

Example 2. Let v and v' be as in **Example 1**. Then

$$3 * v - \{3 \cdot (eur), 15 \cdot (usd), 36 \cdot (eur)\},$$
$$v + v' = \{7 \cdot (eur), 10 \cdot (usd), 24 \cdot (rur)\},$$
$$v - v' = \{\emptyset\},$$
$$v' - v = \{5 \cdot (eur)\},$$

$$v \cup v' = \{6 \cdot (eur), 5 \cdot (usd), 12 \cdot (rur)\},$$
$$v \cap v' = \{1 \cdot (eur), 5 \cdot (usd), 12 \cdot (rur)\}. \blacksquare$$

Common feature of all described operations, which are known from theory of multisets [13, 14], is that their operands are multisets and integer numbers. Unlike them, following operation called *"filtration"* applies set of multisets (*SMS*) as the first operand and so called *"filter"*, being set of *conditions*, as the second operand. Filtration is denoted as $V \downarrow F$, where V is filtrated *SMS*, F is filter, and "$\downarrow$" – symbol of operation.

Conditions entering filter F may be boundary and optimizing.

Boundary conditions (*BC*) are recorded as $a\theta n$, where $\theta \in \{>, <, \geq, \leq, =\}$. Multiset v satisfies *BC* $a\theta n$, if $m \cdot a \in v$ and $m\theta n$ is true (as everywhere, $a \notin v$ is equivalent to $0 \cdot a \in v$).

Let $F_{\leq} = \{bc_1, \ldots, bc_k\}$ be a filter, containing only boundary conditions.

Then

$$V \downarrow F_{\leq} = \bigcap_{i=1}^{k} (V \downarrow \{bc_i\}). \tag{11}$$

Example 3. Let $V = \{v_1, v_2, v_3\}$, where

$$v_1 = \{4 \cdot (eur), 3 \cdot (rur)\},$$
$$v_2 = \{8 \cdot (usd), 10 \cdot (rur)\},$$
$$v_3 = \{3 \cdot (eur), 19 \cdot (usd), 17 \cdot (rur)\},$$

and $F_{\leq} = \{(usd) > 3, (rur) \leq 12\}$. Then

$$V \downarrow F_{\leq} = (V \downarrow \{(usd) > 3\}) \cap (V \downarrow \{(rur) \leq 12\}) = \\ = \{v_2, v_3\} \cap \{v_1, v_2\} = \{v_2\}. \blacksquare$$

Optimizing conditions (*OC*) are recorded as$a = opt$, where $opt \in \{max, min\}$. Multiset $v \in V$ satisfies *OC* $a = max$, if $n \cdot a \in v$ and all multisets $v' \in V - \{v\}$ satisfy boundary condition $a \leq n$. Similarly, multiset $v \in V$ satisfies *OC* $a = min$, if $n \cdot a \in v$ and all multisets $v' \in V - \{v\}$ satisfy boundary condition $a \geq n$.

Let $F_{opt} = \{oc_1, \ldots, oc_l\}$ be filter, containing only optimizing conditions. Then, similarly to (6),

$$V \downarrow F_{opt} = \bigcap_{i=1}^{l} (V \downarrow \{oc_i\}). \tag{12}$$

Example 4. Let V be the same as in Example 3, and $F_{opt} = \{(usd) = max, (rur) = min\}$. Then

$$V \downarrow F_{opt} = (v \downarrow \{(usd) = max\}) \cap (v \downarrow \{(rur) = min\}) = \{v_3\} \cap \{v_1\} = \{\emptyset\}. \blacksquare$$

If filter F contains both boundary and optimizing conditions, i.e.

$$F = F_{\leq} \cup F_{opt}, \tag{13}$$

then

$$V \downarrow F = (V \downarrow F_{\leq}) \downarrow F_{opt}, \tag{14}$$

i.e. result of filtering set of multisets V by filter F is obtained by application of boundary subfilter $F_{\leq}$ to V, after what resulting *SMS* is filtered by optimizing subfilter F_{opt}.

Application of filters, containing boundary and optimizing conditions, is illustrated by **Figure 1**.

Considered operations on multisets and their sets make it possible to define syntax and semantics of family of multiset grammars.

Background of this family is notion of *multiset grammar* as a couple $S = \langle v_0, R \rangle$, where v_0 called *kernel* is multiset, and R called *scheme* is finite set of so called rules. Set of all objects used in kernel and scheme of *MG S* is denoted as A_S.

Rule $r \in R$ is a construction

$$v \to v', \tag{15}$$

where multisets v and v' are called respectively *left part* and *right part* of the rule r, and "→" is divider. The only restriction on left and right parts of the rule is$v \neq \{\emptyset\}$.

If $v \subseteq \boldsymbol{v}$, then result of *application* of rule r to multiset $\boldsymbol{v}$ is multiset

$$\boldsymbol{v}' = \boldsymbol{v} - v + v'. \tag{16}$$

Speaking informally, (16) defines, that if left part of the rule, i.e. multiset v, is included to *MS* $\boldsymbol{v}$, then v is replaced by right part of this rule, i.e. multiset v'. Result of application of rule r to multiset $\boldsymbol{v}$ is denoted as

$$\boldsymbol{v} \overset{r}{\Rightarrow} \boldsymbol{v}', \tag{17}$$

and it is said, that *MS* $\boldsymbol{v}'$ is generated from *MS* $\boldsymbol{v}$ by application of rule r. If left part v is not included to *MS* $\boldsymbol{v}$, result of application r to $\boldsymbol{v}$ is empty *MS* $\{\emptyset\}$.

Set of multisets, generated by application of multigrammar $S = \langle v_0, R \rangle$, or, just the same, *defined by this multigrammar*, is recursively created as follows:

$$V_{(0)} = \{v_0\}, \tag{18}$$

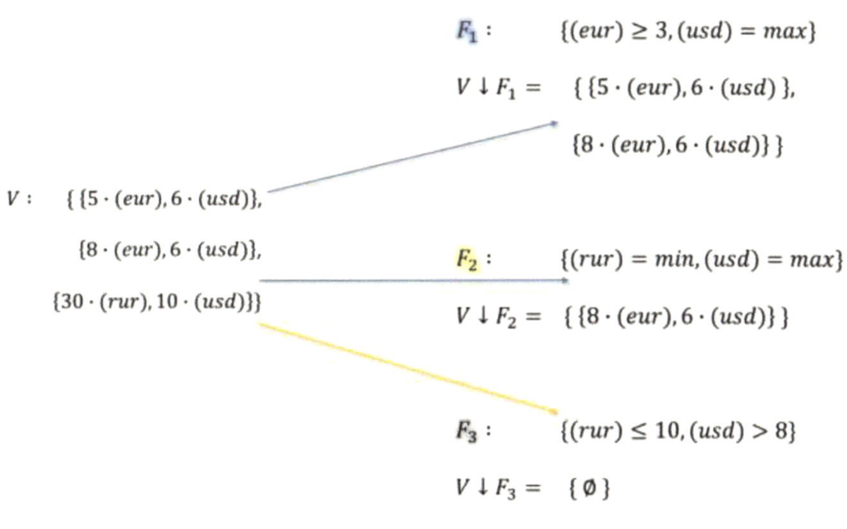

Figure 1.
Filters application.

$$V_{(i+1)} = V_{(i)} \cup \left(\bigcup_{\boldsymbol{v} \in V_{(i)}} \bigcup_{r \in R} \left\{ \boldsymbol{v}' \middle| \boldsymbol{v} \overset{r}{\Rightarrow} \boldsymbol{v}' \right\} \right), \tag{19}$$

$$V_S = V_{(\infty)}. \tag{20}$$

As seen, V_S includes all multisets, which may be generated from $MS\ v_0$ by sequential application of rules $r \in R$, and V_S is fixed point of the sequence $V_{(0)}, V_{(1)}, \ldots, V_{(i)}, \ldots$, so

$$V_S = \bigcup_{i=0}^{\infty} V_{(i)}. \tag{21}$$

In general case V_S may be infinite.

If $MS\ \boldsymbol{v}'$ may be generated from $MS\ \boldsymbol{v}$ by application of some sequence (chain) of rules entering scheme R, it is denoted as

$$\boldsymbol{v} \overset{R}{\Rightarrow} \boldsymbol{v}', \tag{22}$$

and, if so,

$$V_S = \left\{ \boldsymbol{v} \middle| v_0 \overset{R}{\Rightarrow} \boldsymbol{v} \right\}. \tag{23}$$

Multiset $\boldsymbol{v} \in V_S$ is called *terminal multiset* (*TMS*), if

$$(\forall r \in R)\ \boldsymbol{v} \overset{r}{\Rightarrow} \{\emptyset\}, \tag{24}$$

i.e. no any rule $r \in R$ may be applied to this multiset. Set of terminal multisets (*STMS*) defined by multiset grammar S is denoted $\overline{V_S}$. Evidently,

$$\overline{V_S} \subseteq V_S. \tag{25}$$

Example 5. Let $S = \langle v_0, R \rangle$, where kernel

$$v_0 = \{3 \cdot (eur), 6 \cdot (usd), 5 \cdot (rur)\},$$

and scheme $R = \{r_1, r_2\}$, where r_1 is

$$\{2 \cdot (eur)\} \to \{4 \cdot (usd)\},$$

and r_2 is

$$\{2 \cdot (usd), 3 \cdot (rur)\} \to \{2 \cdot (eur)\}.$$

According to (18)–(19),

$$V_{(0)} = \{\{3 \cdot (eur), 6 \cdot (usd), 5 \cdot (rur)\}\},$$
$$V_{(1)} = V_{(0)} \cdot \cup \{\{1 \cdot (eur), 10 \cdot (usd), 5 \cdot (rur)\}\}, \{5 \cdot (eur), 4 \cdot (usd), 2 \cdot (rur)\},$$
$$V_{(2)} = V_{(1)} \cup \{\{3 \cdot (eur), 8 \cdot (usd), 2 \cdot (rur)\}\},$$
$$\ldots$$

As seen, this *MG* provides generation of all possible collections of Euros, US dollars, and Russian rubles, which may be obtained from the initial collection v_0 by sequential currency exchanges, which parameters are fixed by rules r_1 and r_2 (2 Euros may be exchanged to 4 US dollars, 2 US dollars and 3 Russian rubles may be exchanged to 2 Euros).∎

Generalized scheme of application of multiset grammars is presented at **Figure 2**.

Due to the fact, that multiobjects contain both numerical and symbolic components (multiplicities and object names), generation of multisets provides knowledge-driven numerical computation, that creates a lot of new opportunities for simple formalizing and effective solution of various sophisticated practical problems with hard combinatorial background. To implement such opportunities, so called filtering multiset grammars were proposed in [1, 2]. *FMG* are such generalization of *MG*, that integrate two basic concepts—generation of set of multisets and selection from it such *MS*, that satisfy some logical conditions, joined to filter.

Filtering multiset grammar is a triple $S = \langle v_0, R, F \rangle$, where v_0 and R are, as above, kernel and scheme, while F is a filter, including boundary and optimizing conditions, defining multisets, which would be selected from the set of *TMS*, generated by *MG* $\langle v\_0, R \rangle$, i.e.

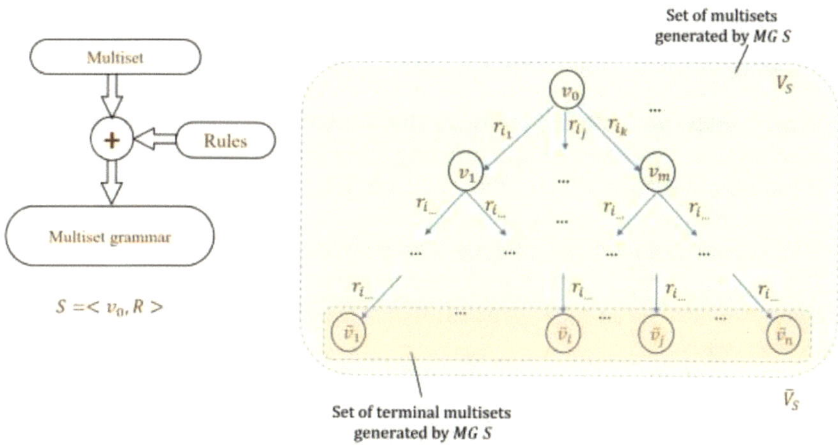

Figure 2.
Application of multiset grammars.

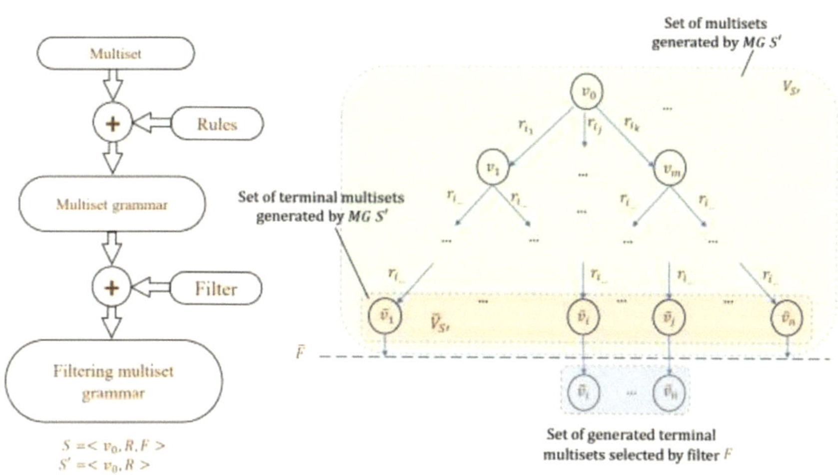

Figure 3.
Application of filtering multiset grammars.

$$V_s = \overline{V}_{\langle v_0, R\rangle} \downarrow F. \tag{26}$$

Verbally, V_s is subset of $\overline{V}_{\langle v_0, R\rangle}$, which includes only such elements of this set, that satisfy filter F. Generalized scheme of application of filtering multiset grammars is presented at **Figure 3**.

After description of syntax and semantics of filtering multiset grammars we may move to their implementation issues, which background are multi-agent technologies.

3. Basic techniques of the multi-agent implementation of multisets generation

Due to granularity and natural internal parallelism of multigrammatical representation it is perspective to try to use multi-agent paradigm as a background for implementation of the *FMG* application engine (*AE*), providing generation of multisets.

We shall use scheme, depicted at **Figure 4**, as a basis for primary multi-agent implementation of *FMG AE*. Proposed multi-agent system implementing *FMG* $S = \langle v_0, R, F\rangle$, contains $l = |R|$ agents $r_1, \ldots, r_l$, each corresponding to one rule from set R (everywhere below we shall denote rules and implementing them agents by the same symbols); one agent F^*, implementing filter F; one agent G^*, providing supervision on other agents interaction (in fact, implementing ubiquitous generation by application of rules to multisets, generated at previous steps). Agent G^* uses storage V_S for accumulation of all generated terminal multisets, satisfying filter F. Also G^* operates storage V, containing current set of generated multisets, which are not yet transferred to other agents. Initial state of V is $\{v_0\}$. Agents communicate via network N.

Described multi-agent system is operating according to the definition of mathematical semantics of *FMG* (15)–(26). Set of messages, which are circulating between the agents, is represented in **Table 1**.

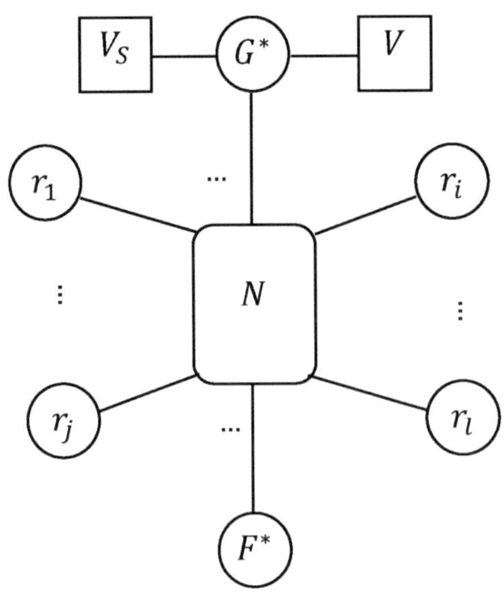

Figure 4.
Multi-agent system implementing FMG.

N	Sender	Receiver	Message	Comment
1	G^*	r_i	$\langle j, v \rangle$	j – number of MS,$v \neq \{\emptyset\}$
2	r_i	G^*	$\langle j, v' \rangle$	v' – result of application of rule r_i to MS v
3	r_i	G^*	$\langle j, \{\emptyset\} \rangle$	Rule r_i is not applicable to v
4	G^*	$\langle j, v \rangle$,	$\langle j, v \rangle$	As 1
5	F^*	G^*	$\langle j, 1 \rangle$	jth MS satisfies filter F
6	F^*	G^*	$\langle j, 0 \rangle$	jth MS does not satisfy filter F

Table 1.
Set of MAS messages.

All messages are couples, the first components of which are numbers of multisets, processed by the agents, and every MS has its unique number. Assigning numbers to multisets is performed by agent G^*. Current maximal number is denoted J. Also agent G^* uses variable Z, which value is set of couples $\langle j, n \rangle$, where n is number of agents r_i which until current moment have not sent to G^* messages with results of application of corresponding rule to jth MS.

Agent G^* sends couple $\langle j, v \rangle$ to all agents r_i. After this, G^* joins to the current value of variable Z couple $\langle j, l \rangle$. Every agent r_i, receiving message $\langle j, v \rangle$ from agent G^*, tries to apply v to rule r_i. If application is possible, r_i sends to G^* message j, v', where v' is result of application of rule r_i to MS v. Otherwise r_i sends to G^* message $\langle j, \{\emptyset\} \rangle$. Agent G^*, receiving message $\langle j, v \rangle$, where $v \neq \{\emptyset\}$, assigns J new value $J + 1$, and sends message $\langle J, v \rangle$ to all agents. Also couple $\langle J, l \rangle$ is joined to Z, and couple $\langle j, q \rangle \in Z$ is eliminated from set Z, that means at least one rule was applied to jth MS, so this multiset is non-terminal. If agent G^* receives message $\langle j, \{\emptyset\} \rangle$ from r_i, that means rule r_i was not applied to jth multiset, and $\langle j, q \rangle \in Z$ is replaced by $\langle j, q - 1 \rangle$. If now $\langle j, 0 \rangle \in Z$, that means no one rule was applied to jth MS, that's why it is terminal, and, according to FMG semantics, it must be filtered. So agent G^* sends couple $\langle j, v \rangle$ to agent F^*, which provides testing, whether v satisfies boundary subfilter $F_{\leq} \subseteq F$. If testing is successful, agent F^* sends to agent G^* message $\langle j, 1 \rangle$, otherwise—message $\langle j, 0 \rangle$. In the case $\langle j, 1 \rangle$ couple $\langle j, v \rangle$ is joined to the current value of variable V_s. After no active agents r_i remain, agent G^* applies optimizing subfilter $F_{opt} \subseteq F$ to the aforementioned current value V_s, eliminating from it all multisets, which do not satisfy F_{opt}. Final value of variable V_s is exactly set of terminal multisets, defined by FMG $S = \langle v_0, R, F \rangle$.

As may be seen, proposed multi-agent system provides generation and filtration of multisets by parallel operation of all agents, entering this MAS.

However, there are some evident bottlenecks, limiting speed of multisets generation. Most essential of such bottlenecks are massive transmissions of multiply repeated sets of multiobjects via MAS communication network. To reduce such transmission it is possible not to send MO to agents r_i such, that corresponding them rules are applicable to multisets, having place in the storage V, and this applicability may be recognized directly by agent G^*. If such opportunity may be implemented, only those agents r_i, which, possibly, may apply corresponding rules to current MS, would receive it. By this measure, traffic on MAS communication network may be reduced sharply.

To implement proposed logics, we shall introduce auxiliary database L, which elements would be couples $\langle a, \{\langle i_1, n_1 \rangle, \dots, \langle i_k, n_k \rangle\} \rangle$, where a is name of object, and in couples $\langle i, n \rangle$, entering the set, which is second component of the couple, integer i is number of the rule, which left part contains multiobject $n \cdot a$.

Example 6. Let scheme $R = \{r_1, r_2\}$, where r_1 is

$$\{9 \cdot (eur)\} \rightarrow \{10 \cdot (usd)\},$$

and r_2 is

$$\{5 \cdot (eur), 3 \cdot (rur)\} \rightarrow \{7 \cdot (usd)\}.$$

Then $L = \{ <(eur), \{ <1,9>, <2,5> \} >, <(rur), \{ <2,3> \} > \}$.∎

Database L has such internal organization, that there exists associative index, providing direct selection of couple $\langle a, w \rangle$ for object name a. To reduce search in the selected list of couples $\langle i, n \rangle$, it may be created as ordered by increase of multiplicities n. Let $v = \{n_1 \cdot a_1, \dots, n_m \cdot a_m\}$ be a current multiset processed by agent G^*, and $Q = \{\langle a_1, \leq n_1 \rangle, \dots, \langle a_m, \leq n_m \rangle\}$ is set of queries to database L, each providing selection of couples $\left\langle a_i, \left\{ j_1^i, \dots, j_{k_i}^i \right\} \right\rangle$, such that $n_p^i \leq n_i$. It is clear, that only in this case rules $r_{j_1^i}, \dots, r_{j_{k_i}^i}$ may have opportunity to be applied to MS v, and, totally, only those rules, in which all multiobjects from their left parts have multiplicities hot greater than those of the same objects having place in v. So there is an evident criterion for selection of rules, which may be applicable to the current multiset v.

Statement 1. Let $\left\{ \langle a_{i_1}, N_1 \rangle, \dots, \left\langle a_{i_p}, N_p \right\rangle \right\}$ be a set, selected from database L by query $Q = \{\langle a_1, \leq n_1 \rangle, \dots, \langle a_m, \leq n_m \rangle\}$, corresponding to multiset $v = \{n_1 \cdot a_1, \dots, n_m \cdot a_m\}$, and $\{n_{j_1} \cdot a_{j_1}, \dots, n_{j_s} \cdot a_{j_s}\}$ be the left part of rule r. Then r may be applicable to v, if

$$\{a_{j_1}, \dots, a_{j_s}\} \subseteq \left\{ a_{j_1}, \dots, a_{j_p} \right\}. \blacksquare \quad (27)$$

As seen, proposed associative organization of set of left parts of rules entering scheme R provides fast selection of sets of rules, which may be applied to the current multiset.

Let us consider further enhancement of *FMG* application engine, based on the multi-agent technology.

First of all, it is evident, that it is not necessary to send all multiobjects of multiset v, to which ruler_i is applicable, to agent r_i, because replacement of left part of this rule by its right part is local operation, regarding in general case relatively small number of multiobjects in the processed multiset v, while all the rest MO remain unchanged. So it is sufficient to send to agent r_i only tuple $\langle f_1 n_{i_1}, \dots, f_t n_{i_t} \rangle$ of multiplicities of objects $a_{i_1}, \dots, a_{i_t}$, in MS v, such that tuple $A = \langle a_{i_1}, \dots, a_{i_t} \rangle$ is ordered lexicographically set of objects, having place in both left and right parts of rule r, and signs f_i before multiplicities n_{i_j} of objects a_{i_j}, having place in left side of rule r, are "–", while all other are "+". Receiving this tuple, agent r provides computation of tuple $\langle \boldsymbol{n}_{i_1}, \dots, \boldsymbol{n}_{i_t} \rangle$, where $\boldsymbol{n_j} = n_j + \left(f_j n_{i_j} \right), j = 1, \dots, t$. (There may be particular case, when some objects do enter both left and right parts but this singularity is simply handled by positioning to the jth place of tuple A number $-n + n'$, where n is multiplicity of object a_{i_j} in the left part, and n'–its multiplicity in the right part of rule r).

Example 7. Let $v = \{5 \cdot (eur), 10 \cdot (usd), 7 \cdot (rur), 18 \cdot (pound)\}$, and r is $\{3 \cdot (eur), 2 \cdot (rur)\} \rightarrow \{5 \cdot (usd)\}$. Because set of lexemes, entering this rule, is ordered lexicographically as $\langle eur, rur, usd, \rangle$ so tuple, sent by agent G^* to agent r, would be $\langle 5, 7, 10, \rangle$ and agent r would sent to agent G^* tuple $\langle 5-3, 7-2, 10+5 \rangle = \langle 2, 5, 15 \rangle$, and, thus, result of $v \overset{r}{\Rightarrow} v'$ would be $v' = \{2 \cdot (eur), 15 \cdot (usd), 5 \cdot (rur),$

$18 \cdot (pound)\}$. As seen, multiplicity of object $(pound)$ is not transferred to agent r, because this object does not enter rule r.∎

Proposed techniques provides further reduction of traffic on *MAS* communication network and, thus, total time of generation of *STMS*, defined by *FMG*.

Application of the described *MAS*-based generation of *STMS* is flexible as it is only possible: due to granularity of multigrammatical knowledge representation local corrections of *FMG* by replacement of one rules by another are easily reflected by corresponding replacement of only concerned agents without touching the other. The same may be done with *FMG* filters. Such flexibility provides the simplest implementation of the most practically useful "what-if" regimes of application of *MG*-centered knowledge-based decision support systems.

4. Conclusion

Proposed techniques of application of multi-agent technology to the high-parallel generation of sets of multisets, defined by filtering multiset grammars, provides essential growth of speed of creation of *STMS*. However, there are some evident ways of further enhancement of *FMG* implementation upon this background:

- development of methods of matching constructed *MAS* to real homogeneous or heterogeneous hardware in such a way that delays, caused by information exchange between agents, would be minimal;

- development of more efficient *MAT* application techniques concerning some more particular cases of *MG* – first of all, filtering context-free multiset grammars as well as filtering unitary *MG* and unitary multiset metagrammars;

- design of special-purpose computing environments suitable for direct implementation of *FMG* and other dialects of *MG* family;

- development of *MAT*-based techniques of implementation of algorithmics of unitary multiset metagrammars as most practically useful tool of the *MG* family.

Acknowledgements

Author is grateful to Prof. Fred Roberts and Prof. Alexey Gvishiani for useful discussions.

Author details

Igor Sheremet
Russian Foundation for Basic Research, Moscow, Russia

*Address all correspondence to: sheremet@rfbr.ru

References

[1] Sheremet IA. Recursive Multisets and their Applications. Moscow: Nauka; 2010. p. 292 (in Russian)

[2] Sheremet IA. Recursive Multisets and their Applications. Berlin: NG Verlag; 2011. p. 249

[3] Sheremet IA. Multiset analysis of consequences of natural disasters impacts on large-scale industrial systems. Data Science Journal. 2018; **17**(4):1-17. DOI: 10.5334/dsj-2018-004

[4] Sheremet I. Multiset-based knowledge representation for the assessment and optimization of large-scale sociotechnical Systems. In: Vizureanu P, editor. Enhanced Expert Systems. London: IntechOpen; 2019. DOI: 10.5772/intechopen.81698. Available from: https://www.intechopen.com/books/enhanced-expert-systems/multiset-based-knowledge-representation-for-the-assessment-and-optimization-of-large-scale-sociotech

[5] Sheremet I. Unitary multiset grammars and metagrammars algorithmics and applications. In: Vizureanu P, editor. Enhanced Expert Systems. London: IntechOpen; 2019. DOI: 10.5772/intechopen.82713. Available from: https://www.intechopen.com/books/enhanced-expert-systems/unitary-multiset-grammars-an-metagrammars-algorithmics-and-application

[6] Sheremet I. Multiset-based assessment of resilience of sociotechnological systems to natural hazards. In: Tiefenbacher J, editor. Natural Hazards—Risks, Exposure, Response, and Resilience. London: IntechOpen; 2019. DOI: 10.5772/intechopen.83508. Available from: https://www.intechopen.com/online-first/multiset-based-assessment-of-resilience-of-sociotechnological-systems-to-natural-hazards

[7] Shoham Y, Leyton-Brown K. Multiagent Systems: Algorithmic, Game-Theoretic, and Logical Foundations. New York: Cambridge University Press; 2009. p. 532

[8] Schatten M, Tomicic I, Duric BO. Multi-agent modeling methods for massively multi-player on-line role-playing games. In: 2015 38th International Convention on Information and Communication Technology, Electronics and Microelectronics (MIPRO). 2015

[9] Waldrop MM. Free agents: Monumentally complex models are gaming out disaster scenarios with millions of simulated people. Science. 2018;**360**(6385):144-147

[10] Sycara KP. Multiagent systems. AI Magazine. 1998;**19**(2):79-92

[11] Yeoh W, Yokoo M. Distributed problem solving. AI Magazine. 2012; **33**(3):53-65

[12] Van der Hoog S. Deep learning in (and of) agent-based models: A prospectus. 2018. ArXiv: 1706.06302

[13] Petrovskiy AB. Spaces of Sets and Multisets. Moscow: Editorial URSS; 2003. p. 248 (in Russian)

[14] Petrovskiy AB. Theory of Measured Sets and Multisets. Moscow: Nauka; 2018. p. 360 (in Russian)